Philadelphia *and the* Development *of* Americanist Archaeology

Philadelphia *and the* Development *of* Americanist Archaeology

Edited by

DON D. FOWLER *and* DAVID R. WILCOX

THE UNIVERSITY OF ALABAMA PRESS
TUSCALOOSA AND LONDON

Typeface: ACaslon

∞
The paper on which this book is printed meets the minimum
requirements of American National Standard for Information
Science–Permanence of Paper for Printed Library Materials,
ANSI Z39.48-1984.

Library of Congress Cataloging-in-Publication Data

Philadelphia and the development of Americanist archaeology / edited by Don D. Fowler and David R. Wilcox.
p. cm.
Papers of the 3rd Gordon R. Willey Symposium on the History of American Archaeology, held in Philadelphia, Pa. in 2000.
Includes bibliographical references (p.) and index.
ISBN 0-8173-1311-7 (cloth : alk. paper) — ISBN 0-8173-1312-5 (pbk. : alk. paper)
1. Anthropology—Pennsylvania—Philadelphia—History—Congresses. 2. Archaeology—Pennsylvania—Philadelphia—History—Congresses. 3. Archaeologists—Pennsylvania—Philadelphia—Biography—Congresses. 4. Archaeology—United States—History—Congresses. 5. Indians of North America—Antiquities—Congresses. 6. University of Pennsylvania. Museum of Archaeology and Anthropology—History. 7. Philadelphia (Pa.)—Intellectual life—Congresses. 8. United States—Antiquities—Congresses. I. Fowler, Don D., 1936– II. Wilcox, David R., 1944– III. Gordon R. Willey Symposium on the History of American Archaeology (3rd : 2000 : Philadelphia, Pa.)
GN17.3.U6 P48 2003
974.8′01—dc21

2003001149

British Library Cataloguing-in-Publication Data available

Contents

Illustrations

Foreword

PHILADELPHIA HAS NOT FARED WELL in the popular imagination. In leafing through a variety of books on famous quotations, I have found that from Mark Twain ("In Boston they ask, How much does he know? In New York, How much is he worth? In Philadelphia, Who were his parents?") to Howard Ogden ("Philadelphia: all the filth and corruption of a big city; all the pettiness and insularity of a small town"); from S. J. Perelman ("Philadelphia, a metropolis sometimes known as the City of Brotherly Love, but more accurately as the City of Bleak November Afternoons") to Stephen Birmingham ("But in Philadelphia, Philadelphians feel, the Right Thing is more natural and more firmly bred in [them] than anywhere else"), and perhaps most famously in the epitaph proposed by W. C. Fields ("On the whole I'd rather be in Philadelphia"), Philadelphia has frequently been mocked for its presumed narrow-mindedness and isolation.

However, in the realm of scholarly leadership, in general, and archaeological advancement, in particular, "the city that loves you back," according to one of its recent marketing slogans, has played crucial but sometimes underappreciated roles in American intellectual history, especially after the Franklin-Jefferson era. As recent books by Bruce Kuklick (*Puritans in Babylon: The Ancient Near East and American Intellectual Life, 1880–1930*) and Steven Conn (*Museums and American Intellectual Life, 1876–1926*), among others, have shown, Philadelphia people and institutions have facilitated the growth of both Old World and New World archaeology.

As the director of one of the institutions that has been an essential part of the fabric of archaeological endeavors in Philadelphia over the past 115 years, I have been delighted to learn new things about the history of the city and even about my own museum from the papers in this volume. I also have been intrigued by the stimulating analyses of some of its most notable archaeological figures, many of whom finally receive the attention herein that they obviously deserve.

In recent years, the Society for American Archaeology's biennial Gordon R. Willey Symposium on the History of American Archaeology has brought much needed attention to the history of archaeology. The 2000 Willey Symposium from which this highly useful volume emanates was no exception, and the excellent chapters in this book clearly advance scholarly understandings of the development of American archaeology in both its intellectual and social contexts. Although Gordon Willey did not live to see the appearance of this significant contribution, I am certain that he would have applauded its publication and the richness of the information and interpretations of the role that Philadelphia played in the history of the discipline.

JEREMY A. SABLOFF
Philadelphia, May 2002

Introduction

Don D. Fowler and David R. Wilcox

When we were asked by the Society for American Archaeology to organize the third Gordon Willey Symposium in the History of Archaeology, we thought first of the venue where the symposium would occur: Philadelphia. Philadelphians have long been actively involved in the organizing, funding, and doing of archaeology. Much of the organizing, funding, and doing has been carried out through, or the results published by, three vital and venerable cultural institutions, the American Philosophical Society (APS), founded in 1743 (Bell 1997, Carter 1993), the Academy of Natural Sciences, founded 1812 (Nolan 1913), and the University Museum of the University of Pennsylvania, founded 1893 (Dyson 1998, Winegrad 1993). The *Oxford English Dictionary* defines archaeology as "ancient history generally; systematic description or study of antiquities," and more narrowly as "the scientific study of the remains and monuments of the prehistoric period," noting that the term was seemingly first used in 1607 but did not come into common usage until after 1851. Prior to that time, "antiquities" was the commonly used term, denoting phenomena studied by "antiquarians." By whatever name, archaeology in the broad sense has been of interest to Philadelphians for two and a half centuries, and many of them, both through and outside the society, academy, and museum, have made signal contributions thereto over time.

The essays in this volume focus on Philadelphians who were concerned with Americanist archaeology, or in older parlance, the "Archaeology of the New World." Even before the founding of the University Museum, there was also widespread interest in classical archaeology of the Old World, broadly construed to include Greece, Rome, Mesopotamia, and Egypt. This interest, and the individuals and institutions through which they worked, are thoroughly described by Dyson (1998), Kuklick (1996), and Winegrad (1993); hence we deliberately limited the symposium to Americanist work.

Americanist Research Agendas

Elsewhere, we have discussed the historical development of Americanist archaeology in terms of changing research agendas centered on a series of questions about the culture histories of Native American peoples (Fowler 2000, Fowler and Wilcox 1999, Wilcox and Fowler 2002). It is useful to summarize that approach here to provide context for the papers in the present volume. The most basic question in Americanist studies, a question extant for over five hundred years, is that of "origins." When Europeans began speculating about the peoples they encountered in the New World in 1492 and after, questions one and two were "When and by what routes did they get here?" Question three was "Who are these people?" That is, "To which populations in the Old World are they related?" Question four was "How can we answer questions one to three?" (Huddleston 1967, Josephy 1991, Milanich and Milbraith 1989, Royal 1992). Five centuries later, those questions are still on the table. As Europeans, and later, Euroamericans, spread across North, Central, and South America in the sixteenth through nineteenth centuries, they encountered a bewildering variety of native peoples, customs, and languages, as well as tens of thousands of archaeological ruins attesting to a long pre-European culture history of unknown duration but obvious great complexity. Questions five, six, and seven were "How can we account for this complex variety of living peoples, languages, and societies?"; "What are the historical relationships between and among the various ruins?"; and "What are the links (if any) between the ruins and the living native peoples?" Five centuries later, those questions remain on the table.

The history of Americanist archaeology/culture history over the past five centuries can be traced through changing theoretical frameworks, methods of investigation, and accepted canons of evidence (Fowler 2000:23–30, 50–71, 79–103, 233–46). Within archaeology, the central methodological question is "How do you get from the distributions of artifacts, ecofacts [plant and animal remains], and geofacts [soils, sediments, minerals], and their relationships on and in the ground, to valid statements about past human behavior within specific theoretical frameworks?" (Hardesty and Fowler 2001:73; see also Butzer 1982, Clarke 1973). Each theoretical framework carries with it agreed-upon research methods and canons of evidence. These structure how the "getting from . . . to" is done, as well as the "validity" of each statement about past human behavior. Americanist archaeology qua culture history has also pursued genetic models, attempting to link archaeological remains with populations of humans (living or dead) and extant or reconstructed languages. Nearly all Americanist attempts to answer the origins and culture history questions have made such linkages.

Early Inquiries

Answers to some or all of our seven questions in the sixteenth and seventeenth centuries were based on speculation derived from the vagaries of biblical, classical, or legendary sources (Huddleston 1967). The development of Enlightenment science brought more structured forms of inquiry and new canons of evidence within a natural science framework. Research agendas for Americanist archaeology began to be formulated in the 1780s by members of the APS. These Enlightenment scholars, especially Benjamin Smith Barton, Peter S. Duponceau, Thomas Jefferson, and Benjamin Rush, had strong anthropological interests within the general Enlightenment concern to develop a "science of man"—anthropology in the broad sense. Anthropology, as it developed within an Enlightenment framework, was—and is—concerned with how humans came to be, how and when they got where they are across the globe, and with their *commensurability*—how and why they are alike and different in their physical makeups, psyches, languages, societies, and cultures (Fowler 2000:15). From the time of Columbus on, the commensurability question was critical: European savants heatedly debated whether Indian peoples were, or were not, fully commensurable with Europeans. If they were judged not to be, that somehow justified, or made easier, conquest, ethnocide, and slavery. The origins issue was critical also since under European's international law, whichever nation's people first arrived in a pagan land, that nation had precedence in staking later claims to ownership and exploitation of that land and its peoples. Thus in the sixteenth and seventeenth centuries, it mattered a great deal if one or another Indian population could be shown to have derived from somewhere in Europe or even Asia, since that might impinge on a later claim by Spain, France, or England (Seed 1995).

By the end of the eighteenth century, both the origins and commensurability issues were still important, but in different frames of reference, using different canons of evidence. Enlightenment science required the application of reason and inductive methods of data collection. Earlier ideas about the origins of the Indians had centered on vague legends about the Lost Tribes of Israel or purported migrations or voyages by one or another European, Asian, or even African group. But in the 1780s, a more scientific data set was required. Scholars in Europe and India were busily demonstrating the existence of, and genetic connections within, various language families and the implications of those connections for culture history. If genetic linguistic connections could be demonstrated between one or more Old World and New World languages, that would be "hard" evidence for determining origins. This approach was taken up by Benjamin Smith Barton (1797, 1809), who attempted a comparative linguistic study of Old and New World languages. It was not successful because he did not have comparable sets of words for each language; sets of "common appelations," as Jefferson (1944) called them. Jefferson circulated vocabulary lists to develop such data sets. Together with fellow APS member Peter Stephen Duponceau and others, he set in motion an Americanist research agenda item pursued for most of the nineteenth century: a general linguistic classification of American Indian languages, culminating in classifications by the Philadelphia patrician and polymath, Daniel Garrison Brinton (1891; see Darnell 1988, and Darnell, and Hinsley, herein) and John Wesley Powell (1892) of the Bureau of Ethnology, Smithsonian Institution.

The commensurability issue took on other dimensions. The great French savant Georges Louis Leclerc, Comte de Buffon, and his sycophant apologist Cornelius de Pauw, argued that New World plants and animals, including the Indians, were weaker, less robust, and less fertile than Old World populations. Jefferson's (1944) *Notes on Virginia,* published in 1784, was written in refutation of Buffon's assertions, as was the Mexican Jesuit scholar Clavigero's (1979 [1787]) *History of Mexico* (see Gerbi 1973).

As the nascent United States began its inexorable Westward expansion and European nations continued their worldwide colonialist/imperialist domination of much of the world, and as both began to cope with issues of slavery and ethnocide, anthropological commensurability took on new meanings. Among the books in Benjamin Smith Barton's library was a copy of the German anatomist Johann Friederich Blumenbach's (1865 [1786–95]) *Anthropological Treatises . . .* (Ewan 1986:321). It was Blumenbach who established the idea that careful metrical analyses of human crania would provide a true "scientific" basis for distinguishing between "races" and that "cranial capacity," expressed in cubic centimeters, provided an index of relative intelligence between and among races. The history of the application and misapplication of craniometry and anthropometry in the service of racism is well-chronicled elsewhere (Barkan 1992, Shipman 1994). Our purpose in noting it has to do with Samuel George Morton, the Philadelphia physician and savant who was a major figure in the development of craniometric studies in relation to culture history (see below).

When Euroamericans began moving beyond the Appalachians into the Ohio and Mississippi valleys, they marveled at the thousands of mysterious burial mounds and earthen structures of the Mound Builders. Who were the Mound Builders, where had they gone, and when? How did they relate to the origins question? Did the ancestors of the historic Indians build mounds, or had some other people(s) done so (Kennedy 1994, Silverberg 1968)? To answer these and related questions about the Indians, the APS issued a circular in 1799 (Jefferson et al. 1799). It called for systematic compilations of linguistic, ethnographic, and historic data on the Indians and the collection of archaeological data including maps, plans, and detailed verbal descriptions. The APS circular, together with Jefferson's *Notes on Virginia,* are regarded as charters for American anthropology (Hallowell 1960:16–18).

There were some immediate replies to the archaeological queries in the circular (e.g., Sargent 1799), as summarized by Benjamin Smith Barton (1799). The circular was reprinted, and replies dribbled in for years (e.g., Turner 1802, H. H. Brackenridge 1818, and C. W. Short and M. D. Plate 1818). The APS continued to be actively involved in reporting Americanist anthropology, archaeology, and linguistics throughout the nineteenth and twentieth centuries (Wissler 1942, Carter 1993, Goddard 1996), as well as providing research support once the society's grants program was initiated.

After 1821, with the opening of the Santa Fe Trail into the province of Nuevo Mexico, Americans began to note the thousands of ruins in what came to be called the American Southwest. Who had built and then abandoned the ruins? Where had the people gone? How did the ancient southwesterners relate to the Mound Builders and/or to the ancient civilizations of Mexico (Fowler 2000:50–70)?

By 1846, when the United States invaded Mexico, the basic research agenda for Americanist archaeology was in place. The first major item was the origins question. It was by then generally agreed that the New World had been peopled from northeast Asia, across the Bering Strait, or possibly along the Aleutian, Alaskan, and British Columbian coasts. The issues of when, and which populations had migrated, remained unclear. The second major item centered on the builders of the eastern mounds and the southwestern ruins. Who were they, and when and where did they go? Were ruins and mounds somehow related? How did either relate to the high civilizations of Mesoamerica? Throughout the nineteenth century, various Philadelphia scholars contributed data and hypotheses toward answers to those questions.

Philadelphia Scholarly Institutions

We have already noted the origin and early and continuing anthropological work of the APS. A second major cultural institution is the Academy of Natural Sciences of Philadelphia, founded in 1812, which continues to the present day. In 1913 the Academy issued a comprehensive 1419-page index (Nolan 1913) of its publications for the first one hundred years. Its two publications series, a *Journal* (appearing intermittently after 1817 and totaling twenty-one volumes by 1913) and *Proceedings* (begun in 1841 and totaling sixty-two volumes by 1913), contain a variety of anthropological articles reflecting changing ethnological and anthropological interests in the United States and Europe in the nineteenth century. For example, there are various articles by Samuel George Morton on human crania, including his initial article, "Observations on a Mode of Ascertaining the Internal Capacity of the Human Cranium" (Morton 1841). Others include George R. Gliddon's (1842) articles on Egyptian human remains. Gliddon was a phrenologist and showman who coauthored the famous polygenist racialist tome *Types of Mankind* with Josiah Nott in 1854 (Nott and Gliddon 1854). Morton, Nott, and Gliddon were at the center of the polygenist versus monogenist battles in the United States in the 1840s and 1850s (see especially Menard 2001, Stanton 1960).

In the 1840s and 1850s, European archaeologists turned their attention to sites they called *kjökkenmöddinger*—kitchen middens—found along rivers or seashores and filled with remains of fish, shellfish, water birds, and artifacts of the peoples who exploited them. A major controversy brewed over the place of the middens in the European prehistoric sequence, widely reported in scholarly and popular venues (Daniel 1976:87–88). Americans reported similar sites in New Jersey and along San Francisco Bay (Ennis 1866, Gabb 1868) to Edward Drinker Cope, the Philadelphia biologist and paleontologist. Ferdinand Vandeveer Hayden (1866), who had an honorary appointment as geologist and who initiated the first of the civilian-led geographical and geological surveys after the Civil War (Goetzmann 1966), reported the famed pipestone quarry in what was then Dakota Territory. The quarry had fascinated Euroamerican explorers since George Catlin, for whom the quarry stone, Catlinite, is named (Ewers 1979). Finally, the academy's *Proceedings* contain some of the first articles written by C. C. Abbott (1863), who monomaniacally pursued New World "paleoliths" for decades (Meltzer, this volume).

Edward Drinker Cope, whose long-running and acrimonious feud with O. C. Marsh is famous in the annals of science, was closely affiliated with the Academy. He bought the journal *American Naturalist* in 1877, which he edited until his death in 1897. The journal's subtitle under the Cope aegis was, *Devoted to the Natural Sciences in Their Widest Sense.* Since nineteenth-century anthropology was seen as a natural science, Cope, as had his predecessor, included anthropological and archaeological articles in the journal, as well as a monthly (sometimes bimonthly) column on the latest developments in anthropology and archaeology. The column was successively conducted by Otis T. Mason and Thomas Wilson, both of the U.S. Natural History Museum, followed in 1894 by Henry Mercer (Conn 1998:160–91) of Philadelphia. Upon Cope's death in 1897 (Frazer 1897), the young anthropologist, Frank Russell, conducted the anthropology column until his own untimely death in 1903. The point here is that Cope, based in Philadelphia, provided a major source of current notes and news about developments in anthropology and archaeology, as well as numerous articles on those topics during the crucial period, 1875–1900, in American history when anthropology was crystalizing into a professional discipline (Darnell 1998, 2001; Hinsley, this volume; Conn, this volume).

Most of the great anthropology and natural history museums in America and Europe were established in the nineteenth century (Barber 1980, Conn 1998, Fowler 2003, Sheets-Pyenson 1988). Both the Field Columbian Museum in Chicago (Wilcox 2002) and the University Museum of the University of Pennsylvania (Kuklick 1896) were created between 1889 and 1893. They were similar in being founded by wealthy and prominent citizens of the two cities, but with very different aspirations. The Field Museum was to be a general natural history museum designed to compete with the American Museum of Natural History in New York (Nash and Feinman 2002), while the University Museum focused on archaeology, primarily Old World (Dyson 1998, Kuklick 1996). Over the past century, the University Museum has grown into one of the great anthropology museums of the world. The story of its origin is discussed in this volume by Hinsley, and Danien and King. Its early, if abortive, efforts to initiate an Americanist program are discussed in this volume by Wilcox (see also Rowe 1954).

The Papers

To sum up, Americanist anthropology and archaeology began toward the end of the eighteenth century. Key roles in the beginnings and the subsequent development of the disciplines were played by Philadelphians and by individuals closely affiliated with Philadelphia cultural institutions, such as Thomas Jefferson. Philadelphians and their cultural institutions played major roles in the development of not only Americanist anthropology and archaeology but world anthropology and archaeology throughout the twentieth century and continue to do so into the new millennium. The papers herein provide further context for understanding the development of Americanist archaeology as well as biographical sketches of a number of individuals prominent in the doing of archaeology and creating the institutions within which most of the work was done.

Curtis Hinsley presents an overview of and context for the development of archaeology in Philadelphia in the crucial years as anthropology and archaeology were in transition toward professionalism. He also provides insight into the often conflicting interpersonal relationships that are part of institution building, in this instance the University Museum. Regna Darnell reviews the roles played by Philadelphia linguist and anthropologist Daniel Garrison Brinton as Americanist anthropology and archaeology became professionalized. A key element in all this was how the fields and their subfields were to be defined. Debates over nomenclature occupied American anthropologists and archaeologists from the 1870s until after 1910. At stake, as in other disciplines (Collins 1998), were different theoretical frameworks and the methodologies and canons of evidence that flow from them. The "winning" framework would dominate Americanist anthropology for decades (Darnell 1998, 2001). Elin Danien and Eleanor King have written a welcome and useful biography of the inimitable Sara Yorke Stevenson, who had everything to do with the development and success of the University Museum.

David Meltzer provides an insightful analysis of C. C. Abbott's attempts to demonstrate the existence of "paleoliths," stone tools similar to those in Middle Pleistocene river gravels in Europe that were thought, by context, to indicate great antiquity. Abbott operated on the general typological assumption that look-alikes are alike. His paleoliths looked like the hand axes being retrieved from the European gravels. If both were the same age, the implications for the origins issue and peopling of the New World were enormous. But look-alikes don't necessarily arise from the same causes and need not be of the same age. And therein lies Abbott's tale.

David Wilcox's chapter on Frank Hamilton Cushing reviews a famous incident in Americanist archaeology: the accusation that Cushing, regarded by many as a genius, but by others as a charlatan, faked artifacts. Cushing's archaeological work in Florida was funded by prominent Philadelphians and sponsored both by the University Museum and the Smithsonian Institution. The incident highlights the fragility of scientific reputations in an era before the process of professionalization had created a firm consensus on the canons of evidence that apply in such cases. A central concern in archaeology is the trustworthiness of data. Charges of fakery or "sloppy" excavation techniques immediately raise the issue of trust and its implication for the individual's reputation and the soundness of her/his scientific claims (Fowler and Salter 2004). Judging where the truth lies is a challenge for all of us.

The chapter by Lawrence Aten and Jerald Milanich describes the work of certainly one of the most colorful early Americanist archaeologists, from Philadelphia or elsewhere. Clarence B. Moore's work on the Mound Builders problem, traveling by steamboat throughout the Southeast for three decades, was a major accomplishment. Most of his results were originally published by the Academy of Natural Sciences of Philadelphia or in Cope's *American Naturalist,* and many are now being republished by the University of Alabama Press.

Frances Joan Mathien chronicles one of the many connections of Philadelphia archaeologists, or archaeological aficionados, with the North American Southwest, in this case Lucy Wilson. While her work was not well known at the time, it was important not only in results but in linking Wilson and the Philadelphia Commercial Museum with Edgar Lee Hewett, founder of the School of American Archaeology (later the School of American Research) in Santa Fe. Mathien brings Lucy Wilson's work into the context of her times and recognizes her contribution to southwestern archaeology.

Finally, Robert Schuyler provides an excellent summary of a part of the long and varied career of a unique and outstanding Americanist archaeologist, John L. Cotter. In his earlier career, Cotter made significant contributions to the origins issue. His 1930s work at the famed Clovis Site in New Mexico, published by the Philadelphia Academy of Natural Sciences, remains as a major contribution to the definition of what many have seen as the culture carried by the first migrants into the New World (Cotter 1937, 1938). Schuyler focuses on Cotter's later career as a very important developer and shaper of American historical archaeology, the archaeology of those who migrated, willingly or unwillingly, to the New World generally after 1492. Here too Cotter's contributions were signal, helping to put historical archaeology on the sound methodological and theoretical basis it now enjoys.

Steven Conn provides a summary and analysis of the papers in the volume from the perspective of a historian of cultural institutions, and Alice B. Kehoe provides a different view from the perspective of an archaeologist and sociologist of knowledge. Both help us realize the importance of the contributions to Americanist archaeology that Philadelphians and their cultural institutions have made for two centuries. We hope that this volume will stimulate even greater interest in the history of Philadelphia archaeology and the institutions developed to pursue archaeological studies. Key factors are the relationships between archaeologists and patrons, whose similar but sometimes conflicting motives reveal why the trajectories of anthropological inquiry have followed certain pathways and not others.

Abbreviations

AAP	Alexander Agassiz Papers, Museum of Comparative Zoology, Harvard University
AIA	Archaeological Institute of America
APS	American Philosophical Society
BAE	Bureau of American Ethnology
BEF	Babylonian Exploration Fund
BMA/CAC	Brooklyn Museum of Art, Culin Archival Collection
CC	Cushing-to-Culin correspondence
CCA/PU	Charles Conrad Abbott Papers, Princeton University
CT/MSP	Cyrus Thomas Papers, Mound Survey Papers, National Anthropological Archives, Smithsonian Institution (Washington, D.C.)
DAB	Dictionary of American Biography
FWP/HU	Frederic Ward Putnam Papers, Harvard University
GFW/OCA	George Frederick Wright Papers, Oberlin College Archives
HCM/BCHS	Henry C. Mercer Papers, Bucks County Historical Society
HFL	Huntington Free Library
HFL/CLB	Huntington Free Library, Cushing Letter Book
HAE	Hemenway Archaeological Expedition
MP	Mercer Papers, Mercer Museum, Doylestown, Pennsylvania
NAA	National Anthropological Archives
PEHC	Peabody-Essex Museum, Hemenway Collection
PHE	Phoebe Hearst Expedition
PMP/HU	Peabody Museum Papers, Harvard University
RDS/UC	Rollin D. Salisbury Papers, University of Chicago
SAR	School of American Research
SHA	Society for Historical Archaeology

SWM	Southwest Museum
UPM	University of Pennsylvania, Museum Archives
WHH/SIA	William H. Holmes Papers, Smithsonian Institution Archives

Philadelphia *and the* Development *of* Americanist Archaeology

1

Drab Doves Take Flight

The Dilemmas of Early Americanist Archaeology in Philadelphia, 1889–1900

Curtis M. Hinsley

In the remarkable observations of his native country published in 1905 as *The American Scene,* Henry James captured the unassuming grace of Philadelphia with the colors of the breast of a dove: Quaker drab mixed delicately with shades of iridescent pink, silver, and green. James's urban, Quakerly dove embodied the "sane society" that he claimed to have found uniquely in Philadelphia, by which he meant to suggest true, consanguineal social relations founded on generations of familiarity and gentility. To be sure, James realized, this deeply rooted world coexisted within a larger, aggressive "pestilent city" of anonymity, anomie, and separateness, but the very distinction—the "Happy Family" versus the "Infernal Machine"—was crucial to James. Especially vital in Philadelphia, as he briefly experienced the fin-de-siècle city, was a tangible American history, not a trumped-up set of memorials and parades but what he came to call Philadelphia's "lucky legacy of the past" (James 1961 [1905]:278–85). Standing in Independence Hall, in the very room of the signing of the Declaration of Independence, the expatriate pondered the quiet persistence of the past in the face of a constantly abrasive, intrusive present.

> The present comes in and stamps about and very stentorously breathes, but its sounds are as naught the next moment. . . . The lapse of time here, extraordinarily, has sprung no leak in the effect; it remains so robust that everything lives again, and we mingle in the business: the old

> ghosts, to our inward sensibility, still make the benches creak as they free their full coat-skirts for sitting down; still make the temperature rise, the pens scratch, the papers flutter, the dust float in the large sun-shafts; we place them as they sit, watch them as they move, hear them as they speak, pity them as they ponder, know them, in fine, from the arch of their eyebrows to the shuffle of their fine shoes [James 1961:293].

Here, in imaginative rendering, James reclaimed the colonial past of Philadelphia from an invasive present—the racial, linguistic, and cultural heterogeneity that had shocked him in Boston, the pushiness and tasteless opulence that had appalled and repelled him in New York.[1] As Lawrence Levine has observed of James's class and generation, such men sought to preserve (or create) "enclaves of culture that functioned as alternatives to the disorderly outside world and represented the standards, if not the total way of life, they believed in" (Levine 1988:177). Remarkably to James, Philadelphia seemed at the turn of the century to have succeeded in perpetuating, into a corrupt and chaotic urban-industrial era, a coherent and genteel leadership class. In *Philadelphia Gentlemen: The Making of a National Upper Class,* his classic study of Philadelphia's social establishment between 1860 and 1940, E. Digby Baltzell (1958:xii) showed how this might in fact have been accomplished: throughout this tumultuous period of American urban growth, he suggested, in Philadelphia "successful merchants, businessmen and bankers, often of lowly origins and crude plutocratic values themselves, produced families whose members, in each successive generation, allied themselves through education, club membership and marriage with families of older wealth and thus served to perpetuate a continuity of upper-class values and authority."

The social and financial elite that persisted and thickened, however, in late-nineteenth-century Philadelphia was a "business aristocracy" that sacrificed the obligations of political and intellectual leadership to material comfort and economic security (Baltzell 1958:xii). Furthermore, a strong tradition of "privatism" in Philadelphia hampered every effort to meet the public needs of a rapidly growing metropolis in the nineteenth century. The city of Philadelphia, Sam Bass Warner concluded in *The Private City,* "was to be an environment for private money-making and its government was to encourage private business." Philadelphia's other tradition of egalitarianism—"an open society where every citizen would have some chance, if not an equal chance, in the race for private wealth"—coexisted unhappily with this powerful tradition of privatism (Warner 1968:99). By the 1890s Philadelphia's Protestant elite had begun to abandon the central city, moving out from fashionable Rittenhouse Square, where they had collected for decades, to the suburbs along the Main Line and out to Chestnut Hill (Baltzell 1964:121–22). Henry James's drab Quaker dove was taking flight.

Philadelphia shared with her sister cities yet other agonies of historical transformation. In the course of the nineteenth century the pre–Civil War vision of the American city as a site of civic conversation gave way, virtually without exception, to cultural realignments in which the university emerged as the primary site of knowledge production and intellectual exchange—at the cost of such antebellum institutions as local libraries, lyceums, and literary or scientific societies. The decades on either side of the Civil War were the period of flux in this transformation; by the last quarter of the century, with the establishment of new graduate programs at Harvard and Johns Hopkins, a new pattern began to emerge: serious discourse about politics, the arts, and the sciences henceforth would take place in association with urban universities; cultural institutions not so connected would struggle for legitimacy under the new conditions, constantly at risk of association with "low-brow" parties and purposes (Orvell 1989:38).[2]

In an era of ascendant commercialism, no other option appeared feasible to those who cared seriously about cultural conversation. As Bostonian Theodore Lyman expressed the situation to his brother-in-law, Alexander Agassiz, in 1873: "Just now there is a tidal stream of commercial life which sweeps into itself all the energy and talent of the United States—only here and there is it resisted by men of peculiar temperament or peculiar genius. . . . What we must keep trying to do—and what we have done very successfully—is to make Harvard College larger and as many sided as possible—that is, to present learning in as many forms as possible" (Lyman to Agassiz, 4 May 1873, AAP).

In trying to understand the struggle to establish North American archaeology in Philadelphia in this period we should bear in mind the larger dynamics of class and culture, which played out with specific results in the Philadelphia context. In the first place, Philadelphia possessed a proud tradition of cultural and educational institutions, beginning with Franklin's American Philosophical Society (1743), the University of Pennsylvania (1740), and Charles Willson Peale's Museum (1786), and continuing in the first half of the nineteenth century with the Pennsylvania Academy of the Fine Arts (1805), the Academy of Natural Sciences (1812), and the Franklin

Institute for the Promotion of the Mechanic Arts (1824), among others. After the Civil War, with the appointment in 1880 of Dr. William Pepper as provost (chief officer) of the University of Pennsylvania, the university began to change from a "small and sleepy place dominated by its well-known medical school" into a "bustling institution of national and international renown" (Kuklick 1996:27). In *Intellect and Public Life,* Thomas Bender (1993) identified the middle third of the nineteenth century as the critical transitional period (as suggested above) for urban public discourse. In Philadelphia the course of this change was difficult because of the strength of existing associations, loyalties, and the privatistic tradition. Ambitious for himself and for the university but aware of the reluctance of the city's wealthy to support university programs directly, during the eighties Pepper established a strategy of indirect support of associated institutions and activities, often remaining deliberately ambiguous about their connections (if any) to the university. In this manner, for example, he promoted in 1887 the Babylonian Exploration Fund (BEF), at the same time creating a program in Semitic languages at the university. Consciously taking advantage of the desire of well-heeled, Protestant Philadelphians to be associated with explorations in the Holy Land, Pepper promised to receive, store, and display all finds from the BEF's expeditions to the Near East. Two years later he built a new library, which became the first repository of the BEF archaeological collections—and thereby the first home of what eventually became the University Museum (Kuklick 1996:27–29; Danien and King, this volume).

To such patrons, it is important to emphasize, the moral and aesthetic gulf between the classical Old World of Rome, Greece, and the biblical Near East on one hand and the barbarous, aboriginal New World on the other seemed as wide as the Atlantic Ocean (Hinsley 1985, 1993; Renfrew 1980). While one must be cautious in reading too much into individual events, it is instructive that at the first annual meeting of the Archaeological Institute of America (AIA) in Boston in 1880, John Wesley Powell argued for support of North American studies in anthropology and archaeology. Prominent Bostonian and AIA founding member Charles Perkins bluntly contradicted him: the Institute should concentrate on Old World sites, he said, so that "we may lay our hand upon something to be placed in our Museums." F. E. Parker added that even if Americans possessed "all the pottery ware, kitchen utensils and tomahawks" of America's indigenous peoples, "it would be no better for us," because such collections did not "improve the people and repay expenditure." With their sights set on the Mediterranean, such men saw no reason to dig and collect "at a point where the civilization was inferior to our own instead of superior." Charles Eliot Norton, founding spirit behind the institute, stated their general position succinctly to Powell: "What we might obtain from the old world is what will tend to increase the standard of our civilization and culture" (AIA, Minutes, 15 May 1880).

Norton and his like-minded colleagues were expressing a vision of aesthetic and moral education that was central to their own sense of social purpose and responsibility; it was one that almost certainly the Philadelphia patrons of the BEF shared, and it was one that permitted only a secondary place for North American archaeology. Indeed, only the finest, perfect products of human genius had primary place in their plans, and "perfection" they tended to define Eurocentrically; by such standards North America was simply barren of value.[3] For his part, Powell later commented caustically on encounters with such men by observing that "our archaeologic institutes, our universities, and our scholars are threshing again the straw of the Orient for the stray grains that may be beaten out, while the sheaves of anthropology are stacked all over this continent; and they have no care for the grain which wastes while they journey beyond the seas" (Powell 1890:652). Frederic W. Putnam, curator and director of the Peabody Museum of Archaeology and Ethnology at Harvard, felt the sting more locally and personally. Trying to describe to Lewis Henry Morgan the importance and attitudes of Norton, Putnam seemed rueful: "To my knowledge he has never been inside the Peabody Museum, and he has not the slightest idea of what I have been doing or am trying to do there. If you can get him interested in the exploration of the remains of the ancient peoples of America you will be doing a good thing, for he is a man of considerable influence in Cambridge and Boston and he would be well backed up" (Powell to Morgan, 31 January 1880).

The situation in Philadelphia was hardly better. When Pepper founded the University Archaeological Association in 1889 as a fund-raising arm of the new University Museum, donations for Mediterranean work far outstripped those for American excavations; those for 1892 were typical of the period: General Fund, $3,135; Egyptian and Mediterranean Fund, $920; American Fund, $50. Philadelphia patrons centered their attention on a legacy of literacy that fell into two parts: American national (especially colonial) history, and classical (especially biblical) western civilization. Preliterate, prehistoric North America simply fell beyond the margins of their interest.

Complicating the conditions set by ideology, class tastes, and the struggle over new institutional alignments—and closely intertwined with them all—

was a set of personalities circulating around Pepper and the early Americanist efforts at the University Museum that may most charitably be described, in retrospect, as mutually ill fitted and antagonistic. The Philadelphia struggles of the 1890s were in part structural and ideological, to be sure, but there is simply no avoiding the conclusion that personal animosities, cross-purposes, and substantially different value systems created serious turmoil within the very small community of Americanists and their sympathizers and enablers. The central figures, in addition to Pepper, were Daniel Garrison Brinton, Sara Stevenson, Stewart Culin, Charles C. Abbott, and Henry C. Mercer.

Charles C. Abbott was a retired physician with a farmstead along the Delaware River near Trenton, New Jersey (Meltzer, this volume). He had received an M. D. degree at the University of Pennsylvania in 1854 but practiced medicine only two years before deciding to support his family by writing popular books and articles on nature—a decision that meant a life of near poverty. Beginning in the 1870s Abbott became convinced that he had discovered evidences of "ancient man" in the gravels of the Delaware River near his home, and for forty years he attached himself, and the fate of his theories, to Putnam and the Peabody Museum. Putnam came to view Abbott as his private New Jersey source, while Abbott saw Putnam as a scientific road out of Trenton; over the years he sent tons of stone artifacts and hundreds of letters to Putnam, establishing a relationship that was alternately obsequious and vituperative. The following passage, from a letter written on a dreary Sunday afternoon in 1878, provides an early indication of Abbott's sense of intellectual dependency: "But what of the future? Mere arrow-head gathering is impotent to suggest a single new thought, and I seem like Othello, to be without an occupation. Surely to go on digging in the gravel will not tell us anything new. . . . If in the course of your thoughts from day to day, in archaeological matters, any new question arises, which you think it possible, I may be able to throw some light upon, by some new style of field-work or otherwise, please let me know. . . . Have pity on me, and send me an idea" (Abbott to Putnam, October 1878).

So it continued until the end of 1889, when Abbott got his long-awaited lucky break. On the basis of his local work and with Putnam's blessing he became the first curator of archaeology at the University Museum. Putnam reminded Pepper:

> Dr. Abbott was for years placed in a very unpleasant position by the non-belief of many persons in the great discovery which he made show-

> ing that man existed in the Delaware valley at a time preceding the deposition of the Trenton gravel. During those years it was my good fortune to be able to help him, and, through the Peabody Museum, to furnish the means for him to pursue his researches. Now that the scientific world gives him the full credit he so richly deserves, and he is offered an honorable position by the University of Pennsylvania, I am filled with happiness for his sake. (Putnam to Pepper, 28 October 1889)

Alas, Abbott's good fortune was short-lived, as the years of loyal service in the field left him completely unprepared for independent judgment and identity. "*You* made it possible for me to become an archaeologist," he told Putnam. "And as I recently wrote to [Henry W.] Henshaw [at the Smithsonian], every jot and tittle of *ideas* as well as specimens, rightfully belongs and certainly shall go to the Peabody Museum" (Abbott to Putnam, 21 June 1889). Abbott's insecurities in his struggle for a separate professional identity in Philadelphia came out in aggression toward his mentor:

> Dr. Pepper, the provost, and Brinton, [Edward D.] Cope, [Robert C.] Lamborn and [Francis C.] McCauley are to meet with *me* next Wednesday, to discuss the better means to adopt in starting this museum. Now, please understand one thing at the outset. I am invited to give my view as to methods of field work &c., and I do not see how I can refuse, as there will be soon an exhaustive survey (archaeological) of [the] vicinity of Philadelphia and they have piles of money to back it. It is work the Peabody Museum could never do, and I can see no reason why they should not be benefited by my advice on some points. Do you see any reason why I should not talk about such a matter? Can it by any possibility be construed as disloyal to Peabody? If so, I should stay away—and miss a splendid dinner. (Abbott to Putnam, 16 October 1889)

Abbott's brief career in Philadelphia was filled with misery and controversy, as he ran afoul of Pepper, Stevenson, Brinton, Culin, and the trustees of the Museum. While some have seen Abbott's plight as frustrated professionalism (Darnell 1970:82), in truth Abbott had next to no notion of what a curatorship involved and spent days in his office doing virtually nothing. At least as early as fall 1892, Pepper had decided to dispense with him, but for lack of an alternative—"truly useful, faithful men are rare," he told Stevenson—he kept Abbott on a bit longer; toward the end of 1893 he noted that Abbott would soon be gone, leaving "nothing . . . but bad memories" (Pepper to Stevenson, 24 February and October 3, 1893, University of Pennsylvania Museum Archives [UPM]). By mid-1894 Abbott was back on his farm in Trenton.

In background, style, and ambitions Henry C. Mercer presented an entirely different situation. An independently wealthy individual of many faces and phases, Mercer, when he was not traveling, lived a relatively isolated life at his home in Doylestown. Local archaeology became his brief but passionate enthusiasm during the nineties, when he succeeded Abbott as curator of American and Prehistoric Archaeology at the University Museum.[4] Pepper originally hoped that Mercer and Abbott would together "lay out a large plan of aggressive work," resulting in "vigorous progress" for what he called "our American field" (Pepper to Mercer, 14 January 1892 [MP]). As it became clear to Pepper, though, that Abbott was not "gaining in the confidence of our colleagues as a thoroughly scientific archaeologist" (Pepper to Mercer, 15 June 1892, Mercer Papers, Mercer Museum, Doylestown, Pennsylvania [MP]), the provost placed more hope in Mercer for building up the museum's American collections: "I appreciate its value—its enormous value—& that of the American field more highly than any other: I realize that the acquisition of objects should be accompanied by a strict scientific account of the exact conditions under which they have been found. . . . You have a clear field. It is a critical opportunity" (Pepper to Mercer, 10 July 1892, MP). At the same time he sympathized with Abbott and Mercer's frustration with the suggestions of others. "As you know very well, I am utterly ignorant of the whole subject, so far as any authority as an expert goes. I am deeply interested in it, I believe in it heartily, and I want to see our Department of Archaeology grow; but I see at once that there are so many different views and so many different ways of looking at it that it needs a great deal of humor, patience and co-operation all around" (Pepper to Mercer, 26 June 1892, MP).

For the next five years—until 1897—Mercer traveled, observed, and collected constantly, from caves in Virginia to the mounds of the Ohio River valley, from the Yucatan peninsula to Montgomery County, Pennsylvania, working partly for the Museum and partly for the Academy of Natural Sciences (Dyke 1989:45–48). In 1897 he abruptly closed his archaeological career by publishing a collection of papers, *Researches Upon the Antiquity of Man in the Delaware Valley and the Eastern United States* (Mercer 1897). At this point Mercer turned his attention, obsessively, to collecting the agricultural and mechanical tools of colonial America, eventually establishing his own, idiosyncratic museum in Doylestown for displaying the collection (Conn 1998). He showed no further interest in the University Museum or, indeed, American archaeology.

In contrast to the abbreviated trajectories of Abbott and Mercer, Stewart Culin began his anthropological career at the University Museum in 1890 and remained there for more than a decade, until disagreements over museum policy and philosophy with Sara Stevenson and some members of the museum board forced his resignation and removal from Philadelphia. In 1903 he accepted the promising position of curator of the new Department of Ethnology at the Brooklyn Museum of Arts and Sciences.

The son of a modestly successful Philadelphia merchant, Culin became fascinated as a young man by the Chinese community of Philadelphia, undertaking a form of self-trained urban participant-observation in his pursuit of surviving Chinese folklore, materia medica, material artifacts, and games in the city environs (Fane 1991:15).[5] These activities and Culin's enthusiastic participation in the various literary, linguistic, and scientific societies of Philadelphia brought him to the attention of Brinton, who became in effect his sponsor and mentor. In 1890 Brinton and Culin were both appointed to the board of managers of Pepper's new museum; while these were honorary positions, two years later, with the strong support of Brinton, Culin was named the first director of the University Museum (Fane 1991:15–16).

Almost as soon as he joined the museum in 1892, however, a division began to emerge in the pattern of Culin's professional life. In a whirlwind of energy and activity over the next two years, he designed and set up exhibits for the Chicago World's Fair, traveled to Spain to represent and collect for the museum at the Columbian exposition in Madrid, and established the vitally important intellectual camaraderie with Cushing that would lead fifteen years later to his classic study, *Games of the North American Indians* (Culin 1907). But as Culin's national reputation grew, his position at the University Museum became increasingly contentious and even precarious. The problem was a mixture of philosophy, structure, and personality. In the first place, while he bore the title of director, Culin actually had limited oversight of the museum: he was in charge of the physical facilities but had curatorial control only of the departments of American and General Ethnology (Fane 1991:308 fn. 15). The remainder of the departments, and indeed the general supervision and direction of the Museum, seem to have been monitored closely by Stevenson and Pepper; Culin's efforts to extend his directorship over Stevenson's Babylonian, Egyptian, and Mediterranean sections were resisted and resented (Darnell 1970:82; Danien and King, this volume). Furthermore Culin, who was in the process of leaving the business inherited from his father in order to devote himself completely to anthropology and the museum, required an income for his directorship—which was an unusual request and made him, with Abbott, one of the few paid employees of the museum. Contemporary correspondence among Pepper, Stevenson, and members of the board indicate that this requirement alone, in their eyes, placed Culin in an inferior position in the local social hierarchy; that they felt they were simultaneously educating and supporting him; and that, additionally, his abiding interest in Philadelphia's local Chinese folk practices struck them as a matter of private amusement. There is some risk in reading too much into only occasional references, but readers may judge for themselves from the following passages:

> I sincerely trust that his rapid promotion has not turned his head. Unless he is modest and conciliatory and respects the rights of others, we shall be in hot water from morning to night with him. I know your influence with him and I hope you will begin to exert it before he comes back, so that he may step into his position with a proper line of action already mapped out. Please write plainly to him upon this subject. We have a splendid start, and nothing can hinder us but internal dissensions [Pepper to Stevenson, 17 October 1892, UPM].

> But does that wretched Culin work with energy & in accord with our plans? Are the books & collections brought by him from Madrid really precious & important? Today he came to beg me to advance him money. I already had sent him $100 in Spain. I don't want to advance any more until he shows us his collections. He is at least 1/2 an Oriental & with Orientals one must be one & a half Oriental [Pepper to Stevenson, 24 February 1893, UPM].

> Culin was here today. I scolded him to stimulate his activity [Pepper to Stevenson, 9 March 1893, UPM].

> As to Culin, I should like him to go—but it seems to me impossible for a young concern to educate a man & at the same time call him

'Director'—& pay big salary [Pepper to Stevenson, 5 October 1893, UPM].

Owing to your almost continuous absence from the museum since your appointment no special blame can be attached to you for this condition of affairs in the past. I hope for a better management in the future. [Board member C. Howard Colket to Culin, 11 January 1894, regarding rusting of some artifacts].[6]

Culin might have had a better chance to push forward his plans for museum anthropology and Americanist work at Penn had his mentor, Daniel Brinton, stood more firmly behind him. Indeed, it appears that it was Brinton's refusal to lend a strong hand, and take strong stands, within the University Museum circle during these years that may have been the single largest factor in the turmoil and false starts that characterized the early years of Philadelphia anthropology. In the last two decades of the nineteenth century no anthropologist had greater standing in Philadelphia than Brinton. At the founding of the museum in 1889 the good doctor from Media, just over the city line, was entering his last decade of life. Since 1886 he had held a professorship of anthropology in name but not in function at the University of Pennsylvania and enjoyed an international reputation as a scholar of American Indian ethnology, linguistics, and, to a lesser degree, archaeology. Walt Whitman, viewing Brinton from across the river in Camden, New Jersey, was unequivocal in his estimation of Brinton as America's anthropologist:

I think now is the time for archaeology to be exploited here anyhow—especially American archaeology. I remember that when Lord Houghton, Moncton Milnes, called to see me years ago, the first thing he said to me was: "your people don't think enough of themselves; are not in the good sense patriotic enough; they do not realize that they not only have a present but a past, the traces of which are rapidly slipping away from them." He referred to the slack interest we show in remains. We have our schools and expeditions for Greek exploration: the people concerned are begging, begging, all the time for money—which is all right, as far as it goes. I would not put a straw in the way of this—not a straw; I wish it well: it is important work. But I say, why not give our own evidences a chance to show themselves, too? Why not open up our own past—exploit the American contribution to this important science?

> Brinton is doing just that—he is eminent: he insists upon the work and does his part [Traubel 1961:128–29].

Whitman in fact took great interest in Brinton's movements: "Brinton is the best of them all: the first person I always think of in connection with that science [anthropology]," he stated on another occasion in 1888 (Traubel 1961:321).

Despite Brinton's fame, the following year Pepper advised against inviting him to the organizing dinner for his new Archaeological Association, on the grounds of Brinton's well-known hostility toward Edward Drinker Cope, the famous paleontologist and fellow Philadelphian who would also be attending the dinner. (Pepper originally intended that the University Museum have two disciplinary strengths: archaeology and paleontology—but paleontology had generally faded as a field in the museum by 1900.) In the end, both Cope and Brinton did attend and assume leading positions in the new association, with Brinton in charge of raising funds for research efforts in the American hemisphere. Robert C. Lamborn, a key financial supporter of Pepper, expressed his pleasure at the prospect of Brinton's participation: "Philadelphia," he predicted, "bids fair now to resume her old position as the most important centre of scientific growth in America" (Lamborn to Macauley 1889).

Brinton lost no time in issuing a circular calling for subscriptions for six thousand dollars for American work: "It is especially desirable to begin explorations at once in certain of the Archaeological fields within the United States. An opportunity for these is just now offered of unequaled richness, promising the most gratifying returns both in specimens and information concerning the vanished races which previously occupied our territory" (AAUP Circular 1890). Brinton had two fields in mind: Arizona and Florida. Attempting to take immediate advantage of the collapse of Cushing's Hemenway Southwestern Archaeological Expedition, he made overtures directly to both Jesse Walter Fewkes (recently appointed to succeed Cushing) and Adolphe Bandelier (southwestern historian on the Hemenway Expedition). Brinton also apparently intended to employ Clarence B. Moore for archaeological reconnaissance in Florida—a job eventually taken up, ironically, by Cushing in 1895.[7]

Always on the lookout for financial support, Bandelier responded enthusiastically (but secretively) to Brinton's initial appeal; but the subscriptions for American work did not materialize—an unfortunate harbinger of the coming decade, when virtually every effort to bring a scholar of note (including Franz Boas in 1893—see below) into working relationship with the association or the museum failed. The single exception was Cushing's Key Marco work of 1895–96 (Wilcox, this volume).

After Brinton's initial, hopeful appeals of the first months, internal dissensions, founded in personalities and divergent views of museum anthropology, began to sap his enthusiasm for the museum and his relationships with Pepper and Stevenson. Much of Brinton's displeasure apparently centered on Sara Stevenson's influence in the museum; the deterioration was already well advanced when, in early 1893, she asked Brinton directly for a fund-raising commitment and a personal donation to the University Museum:

> As Vice-President of our Dept. and as President or Officer of nearly all the learned bodies in town—[and] as head of the American Section of the Museum, the moral support which your endorsement & cooperation would lend me must be *immense*. . . . As a woman, an enthusiast, I need such endorsement from the authorities from the University, and from those who, *like yourself*, are identified with the archaeological interest in this city, & indeed in this country. . . . [B]ecause it is imperative that work should be done, and that my assumption of this task is not the rash act of an excitable woman, but a simple duty *deliberately* performed by me for the benefit of the University and for the promotion of the science [of archaeology] [Stevenson to Brinton, 28 January 1893, UPM].

Stevenson's self-description as a mere female "enthusiast" in need of official male "endorsement" angered Brinton, raising his own sense of class and professional resentment:

> In some respects, I thoroughly believe in the philosophy of [August] Comte. He taught that society should be divided into two great classes—first, those men and women who are willing to pass their time in the study of science, and for that object renounce the ambitions of practical life; and second, the money-makers, the producers, the workers in applied science; and from the latter should come the support of the former. It is the duty of the rich, the prosperous, the practical citizens of Phila.[delphia] to support our institution. Surely were I one of them, I should aid; but years ago, I deliberately left a profitable business that I might, on a modest competence, pursue my life as an observer, a thinker

> and an unpaid writer. I cannot assume this obligation without forgetting that object; and I am sure you would not wish me to do that. With warm congratulations for your success [Brinton to Stevenson, 30 January 1893, UPM].

Stung by Brinton's sarcasm, Stevenson insisted on having the last word, and thereby revealing her own sense of place—or displacement—in their local struggle over the direction of the museum and its archaeology. She also demonstrated a dismissive attitude toward Brinton and an equal capacity for insult:

> I regret that you do not feel that ardent interest in our undertaking that is arousing me to unwonted energy. But, of course, you are the best judge of your own limitations and it is not for me, I am sure, to urge upon you a distasteful & irksome task. . . . I am so profoundly impressed myself with the importance of the work, that I am almost inclined to regard it as a privilege—that I should have been given a share in it, small though it may be. And in reading your reference to Comte's division of humanity it gives me a shudder to think that, according to your views I have no "raison d'etre"—no place in the world, for I *cannot* remain in the first category, and *do not* belong in the second.
>
> However that may be, one thing is certain: You are a great credit to us, & I am very proud of you! Cold-blooded as you are, you are very nice indeed & I am very fond of you—so [I] forgive you for not "enthusing" with us! Will you think seriously where you can classify poor me? And not leave me out in the cold? Outside the pale of Comte's and y[ou]r philosophy? I feel so lonely, now! [Stevenson to Brinton, 31 January 1893, UPM].

Matters only got worse over the next eighteen months. Brinton objected to proposed arrangements that would subject the University Museum to the operations and decisions of other institutions—notably the Harvard Peabody Museum—and he argued strongly for curatorial freedom to study and publish about collections given to the museum by patrons and other individuals. "A University Museum [he instructed Stevenson] has two main purposes—the one, of investigation, the other, of didactic instruction. That it should be made an attractive showroom, or a sales-room for those with collections to dispose of, is to me unwelcome" (Brinton to Stevenson, 11 July 1894, UPM). And most emphatically, he protested Stevenson's proposal to expand the purview of the Museum beyond anthropology to embrace "artistic and industrial fields." In withdrawing from the Museum board in mid-1894, Brinton summarized his situation:

> You know, of late, your gods have not been my gods, and I have not been able to sympathize cordially with the plans you outline and the methods you favor. . . . I might have to oppose you, and that I do not wish to do any more. Moreoever, I am in certain main points a valueless auxiliary; I have no money to give, and no capacity for raising any—never did have, for anything. As I pursue my scientific studies solely for the pleasure they give me, it has presented itself quite forcibly to me several times in the last six months, that it would be to introduce a fly into my pot of ointment to put myself in a position where I either have to antagonize others, or tacitly consent to what I do not like.
>
> For these reasons I have decided to withdraw from the field. Do not think that I am "disgruntled." I am not that. I have no "grievance." It is simply a deliberate conclusion as to what is most agreeable to myself. You may call it "selfish" if you please, & I shall find no fault. Did we not once have a talk in which we agreed that *all* motives are at heart selfish?
>
> I have no thought of "resigning," but I do not contemplate a reappointment in connection with the Museum. It is not likely that I shall be wanted, as I do not see that I could be of any use, & others will no doubt agree with that view [Brinton to Stevenson, 8 July 1894, UPM].

Eavesdropping on this exchange between the most respected anthropologist in Philadelphia and the most influential woman in the University Museum moves us closer to the conditions that frustrated Americanist archaeology in Henry James's city of the dove. The frustrations had much to do with Stevenson's position and sources of power, and her controlling ambitions, combined with Brinton's withdrawal.

While Brinton had spread the interests and energies of a lifetime over the range of Philadelphia's intellectual societies (a strategy imitated by Culin), Stevenson dedicated herself, for a time, to Pepper's vision for the Museum, invested personally and deeply in it, and thereby gained great influence over it. By most accounts Sara Stevenson (1847–1921) was a determined and intelligent woman, a remarkably successful fund-raiser and social influence, and a competent curator of Egyptian artifacts as well (Darnell 1970:83; Kuklick 1996:60–63; Jacknis 2000:6–7; Danien and King, this volume). She had been born in Paris of American parents and raised in France and Mexico. In her early twenties she settled in Philadelphia and married Cornelius Stevenson, a wealthy lawyer. During the 1880s she joined with prominent Philadelphia gentlemen in their enthusiasm for Egyptian antiquities and Egyptology and soon became a major force in the Babylonian Exploration Fund, Pepper's American Exploration Society, the University Museum's support for Egyptian, Mediterranean, and Near Eastern archaeology, and promotion of the University Museum (Kuklick 1998:61–62). She came to the attention of the national anthropological community through her services at the Chicago World's Fair in 1893 under the directorship of Putnam. Furthermore, through her close friendships with Zelia Nuttall and Phoebe Hearst—who shared both her outlook as a socially prominent woman and the cosmopolitanism derived from time spent in Mexico and Europe—Stevenson brought a global perspective to the world of museum patronage in Philadelphia (Jacknis 2000:6–9). Darnell has suggested that she was "more interested in the material artifacts of exotic cultures than in the broad panorama of human history" that concerned more professionally oriented figures such as Brinton (Darnell 1970:83). If "exotic" is understood to mean Old World, Mediterranean civilizations, this judgment is essentially accurate—and equally applicable to most members of the University Museum board. But it was her close relationship with Pepper, from at least 1889 until his sudden death in 1898, that gave particular force to her presence in the museum.

William Pepper, whose wife was an invalid and who had a reputation as an adulterer and "charlatan," clearly had great attraction for Stevenson, and at his unexpected death in 1898 she endured a serious psychological crisis (Kuklick 1996:61). But Stevenson was also a brilliant aid to Pepper in fund-raising for the museum, and he depended on her for organization, ideas, and constant emotional support. They came to see the museum as their joint project and common fate: "Never will you give up your quest for a Museum, and never shall I give up helping you according to my feeble powers," he assured her. "I should look upon you as false, damned & lost forever did you forsake your great work. I am sure that God will send us some fat subscriptions & then we shall toast each other in a magnum of champagne" (Pepper to Stevenson, spring [n.d.], 1893, UPM). Frequently comparing their progress to Putnam in Cambridge, they buoyed each other at low times: "I am so disgusted with the lack of intelligent interest among our people [in Philadelphia]," he wrote. "But no, that is not fair—it is only because such interest in things scientific & artistic here is too young & too limited to justify our expectations of any spontaneous movement. It will all come. We can wait" (Pepper to Stevenson, 24 February 1893, UPM). In the meantime, they treasured their late-night communications ("These long silent hours of the night are our only refuge & solace"—[Pepper to Stevenson, 4 July 1893, UPM]) and their rare times alone: "Dear Friend—Can it be but a week since we were in the Plaisance [of the Chicago World's Fair]—among the Javanese—just turning the tip top point of the Ferris Wheel—it all seems a part of another life. . . . come home before long. But stay—Stay while results can be accomplished by you" (Pepper to Stevenson, 24 September 1893, UPM).

The Chicago Fair was a great hope and opportunity for Stevenson in particular, since her presence for several months at the Exposition gave the Philadelphia group entree to a national community and contact with individuals such as Putnam, Franz Boas, George Dorsey, and the Washington anthropologists. Stevenson seems to have realized (and convinced Pepper) that Boas, in particular, represented the scientific future of anthropology; at her insistence Pepper made serious efforts to arrange a joint appointment for him between the Wistar Institute and the University Museum. He reported hopefully on progress on Boas and obtaining collections from the Fair:

> Dr. Boaz [*sic*] is free—He has left Clark University—He is open to our offer unless Putnam has secured him for the Columbian Museum. We might have had *him* instead of Abbott or Culin—Great God—to think of it. Can you find out insidiously from Putnam if Boaz is to stay in Chicago? How clever they are in Chicago to make Putnam the head of the Columbian Museum [Department of Ethnology]. They deserve all they will get—for they have been great and brave & wise—but their rewards will be fine. I hear that the U.S. Govt. will not oppose their claims to exhibits. But I believe nothing—except that we need all we can get & that you are more clever than any of them—that we shall do better than any save Chicago & Harvard & that you are the best fellow I ever knew and I wish to Heaven I were in the Midway at this moment with you.

> I saw Wistar this a.m. re Boaz. He is willing to make any arrangement I approve within the limits of the money available. Find out what you can [Pepper to Stevenson, 24 September 1893, UPM].

In the end Pepper was not able to raise the three-thousand-dollar salary that Boas demanded; further, upon her return to Philadelphia at the end of the fair, he was unwilling to move Stevenson from honorary to faculty status at the university. It was a slight that she attributed to expediency rather than prejudice, but she could never forgive him. While she continued to work endlessly to establish a national stature for the museum and to "delocalize" Philadelphia's archaeological efforts (Stevenson to Hearst, 9 October 1898), the intensity of their relationship cooled permanently. She came to depend increasingly through the coming years on her many close female friends, including Hearst, Nuttall, Alice Fletcher, and women of Philadelphia. "My dear Emmie, Do you remember the day we were driving out to the Museum, and you told me of Frank Macauley's death? After passing the Walnut Street bridge, where there is a long fence, we saw a man engaged tearing off the posters in large quantities, and sending them flying all over the city; you were so indignant that you wanted to stop your carriage and harangue the man, and I held you back, saying 'not to-day; we are not in a state of mind to enter into discussion with a man'" (Stevenson to "Emmie," May 1896).

As always, Henry James was perceptive in recognizing a "consanguineal" world of restraint and gentility at the core of urban Philadelphia at century's end—a female state of mind while crossing the Walnut Street bridge. The dilemmas confronting Americanist archaeology in each major urban center of the time—Chicago, Boston, Philadelphia, Washington, New York—were similar overall but distinctive in emphasis and focus: the ambitions and tastes of business elites; the unstable professional world of archaeology (and, more largely, anthropology and the social sciences); the difficult transition to the university as a new arena of civic discourse and knowledge production; and the especially ambiguous role of museums in an urbanizing social order. All these factors were at play in Philadelphia, colored by local conditions and personalities. Brinton, his career in its final stage, possessed enormous prestige but little understanding of how to institutionalize a discipline, either in the museum or in the university, and he completely failed to pass on his particular talents to a next generation. Perhaps he didn't care. In any case, he was hardly unique in this respect; the remarkable counterfigures of his time—the really exceptional men—were Putnam in Cambridge (and New York, Chicago, and Berkeley) and, slightly later, of course, his protégé Boas. Brinton's posture was toward the past and a world of gentlemen's societies that was being rapidly overtaken; Putnam, who also came from that trans–Civil War generation, saw beyond and struggled mightily to transcend it. Mercer and Abbott were simply too idiosyncratic to have any staying power in a large-scale institution, and Culin—who later pursued a successful collecting and curatorial/artistic career at the Brooklyn Museum—was too closely identified as a local, self-educated Philadelphia boy to have a serious future in his native city. At the center of this local stage stood Pepper and Stevenson. They performed, for a brief time, as a formidable duo in the service of the intellectual life of the city as they conceived it, fully convinced that they were "working along lines that will long, long hereafter yield good results—by the genuine amelioration of the masses" (Pepper to Stevenson, 9 July 1893, UPM). Their concern was not so much a matter of profession as a vision of privilege, not so much knowledge production as social obligation. It remained for such figures of the next generation as George Byron Gordon to make the fine distinctions and gentle compromises befitting a new order of professional archaeology.

Notes

1. James's disgust at the changes—the "profane overhauling"—of late Gilded Age America is well known; for a particularly effective and relevant account of his revulsion, see Lawrence W. Levine (1988:171–73).

2. Orvell (1989:38) quotes Edwin L. Godkin's dismissal of the debasements of American culture, which "diffused through the community a kind of smattering of all sorts of knowledge, a taste for 'art'—that is, a desire to see and own pictures—which, taken together, pass with a large body of slenderly-equipped persons as 'culture.'" Godkin, Charles Eliot Norton, and other arbiters of taste agreed that the university must provide the antidote and final defense against such vulgarity.

3. Developing a strategy for AIA support in the western states somewhat later (1906), Francis Kelsey, corresponding secretary and later president of the AIA, observed as follows to Yale professor Thomas Day Seymour: "In Colorado, Utah, New Mexico and other states of the West, there is a large body of college men from the East who cherish the classical traditions. Naturally enough the first impulse of the people in this region is along the lines of immediate [local archaeological] interest; but I am convinced that the interest

in American Archaeology, which is a field almost barren of the ideal, will lead to the development of an interest along the lines of the Institute's present work in classical and Oriental lands" (Kelsey to Seymour, 15 May 1906, AIA Archives).

4. On Mercer's early career see Williams (1991:118–22).

5. There are suggestive parallels between Culin's informal education and early participation in an exotic community and those of Frank Hamilton Cushing, who became his most important intellectual and spiritual guide between 1893 and the latter's untimely death in 1900. It seems clear, for instance, that Culin's early fieldwork and collecting in the Southwest for the University Museum (1901–2) was part personal pilgrimage and part fulfillment of the Cushing legacy. The remarkable correspondence of the 1890s between Culin and Cushing has been collected and is being edited for publication by David R. Wilcox.

6. "The tone of this letter is typical of a series of letters from Colket to Culin from 2/7/93 to 10/5/94, which I have destroyed as having no permanent value. As they dealt (most critically) with all phases of Culin's work, there is an implication that Colket was the Board member in charge [of overseeing Culin's work] until his resignation in March 1894." (Note attached to the Colket/Culin letter by Geraldine Bruckner, first registrar and archivist of the University Museum [1930–64]). Alex Pezzati, the current archivist, suggested that Bruckner was embarrassed by Colket's outspoken comments in his letters about certain individuals, including Sara Stevenson, and so destroyed the correspondence.

7. As Stephen Williams has properly noted (1991:108), Brinton's early archaeological fieldwork (1856–59) was in the shell-heaps of Florida, even before Jeffries Wyman's better-known pioneering work.

2

Toward Consensus on the Scope of Anthropology

Daniel Garrison Brinton and the View from Philadelphia

Regna Darnell

Late-nineteenth-century synergy between local scientific societies in major eastern cities and professionalization of American science at the national level occurred across the disciplines of the social and natural sciences. In anthropology, the newest of the social sciences, the trend toward professionalization played out in the very definition and scope of the emerging discipline. A four-subfield structure has been taken for granted in North American anthropology since the early-twentieth-century coalescence of the historical particularist paradigm around the work of Franz Boas and his students (Darnell 1998).

In retrospect, there was considerable continuity in the emergence of an American mainstream disciplinary paradigm including ethnology, linguistics, archaeology, and physical anthropology. This gradual process, however, has been eclipsed in professional memory, with the logic of disciplinary organization attributed to the German *volksgeist* tradition imported to North America by Boas in the 1880s (Stocking 1995). Boas grafted the anthropology we have inherited today onto a solid preexisting rootstock of intellectual approach, scholarly network, and emerging institutional framework in government and museum research, both emphasizing public dissemination of the results of anthropological inquiry. An established indigenous Americanist tradition preceded Boas (although, of course, he modified it and redirected it toward university-based credentialing, especially after his appointment at

Columbia in 1897). The stakes extended beyond mere labels. When terminology for the discipline was debated, sometimes acrimoniously and always with implications for control of the emerging profession, Boas was not yet a major player in North American anthropology. The initial trajectory of professionalization arose from vested interests and futuristic visions of a coterie of anthropologists with well-established careers by the late nineteenth century (especially Frederic Ward Putnam at Harvard and the Peabody Museum).

Further, the terminological debate was not focused in sociocultural anthropology, as questions of scope have tended to be during the Boasian and post-Boasian phases of Americanist anthropology. Material culture, and by extension archaeology, was particularly salient in the period transitional to professionalization because museums then were the primary supporters of anthropological research. Even in Washington, where the Bureau of American Ethnology dominated anthropological efforts, the United States National Museum tempered founder John Wesley Powell's ethnological and linguistic interests with a healthy dose of archaeology. Although National Museum curator Otis T. Mason was committed to a universal evolutionary framework for the anthropological study of culture, the presence of the Bureau, with its congressional mandate to study the American Indian with a view to scientific determination of public policy, in practice enjoined a national anthropology focused around the study of indigenous peoples.

Powell, like Boas, preferred museum arrangement by culture area and functional use of artifacts. A. Irving Hallowell (1960) argued persuasively that the unity of American anthropology rested in this American Indian mandate, which persisted until after World War II; the quadratic structure was a natural entailment. By implication, other internationally significant national traditions grew from former colonial empires (Britain, France, and Germany) without continuing indigenous traditions central to the national imaginary (Anderson 1983).

In a contemporary context, access to the circumstances and attitudes of late-nineteenth-century North American anthropology requires considerable historicist contextual reconstruction. The transitional period of emerging professionalization was dominated by local scholarly societies and museums in a handful of American cities. In anthropology, three stand out, each with its particular constellation of an anthropological entrepreneur and his institutional resources.

In Washington, D.C., Major Powell ran his anthropological research organization with firm control over group projects designed to document and systematize a database relating to the various Indian tribes and languages (Darnell 1998, Hinsley 1981). Archaeology was a secondary focus but always overlapped with cultural and linguistic work. Powell's institutional framework uniquely facilitated long-term collaborative projects. A linguistic classification filling all gaps in the map of aboriginal North America appeared under his name in 1892 and a synonymy and brief cultural description, the *Handbook of American Indians* edited by F. W. Hodge, in 1907. Later, under the editorship of Boas in his incarnation of honorary philologist to the bureau, the *Handbook of American Indian Languages* appeared in 1911 and 1922 (already drawing on the fruits of three decades of accelerating professionalization, e.g., by incorporating grammatical sketches and texts by Boas's first generation of doctoral students from Columbia University).

The bureau provided critical mass for the professionalization of North American anthropology around the study of the American Indian. At its founding in 1879 Powell's staff took their places among the first anthropologists to earn a living from their professional work. The early staff all had careers in other fields before turning to American Indian ethnology and linguistics under Powell's charismatic leadership. There were no academic programs in anthropology in these years, so their absence of formal training imposed no particular disadvantage. The U.S. Government Printing Office facilitated a substantial publication program of bulletins and annual reports, ostensibly geared to a general (educated) American public rather than exclusively to scientists, whether in anthropology or related disciplines. The practical result set a standard for published and peer-reviewed research reports within anthropology.

The bureau was alone among American institutions, however, in its claim to national scope (although the Peabody Museum may well have been more influential in archaeology) based on its funding by Congress on behalf of the American public. The Smithsonian Institution, in which it was housed, also held a commitment to public dissemination of utilitarian scientific results. Despite the prominence of the Bureau, however, Washington anthropology was institutionally decentralized. The U.S. National Museum and the Anthropological Society of Washington (with its spin-off Women's Anthropological Society of Washington) provided a forum for anthropological debate and publication along the lines of local scholarly societies in other major American cities. The Anthropological Society of Washington membership included many government scientists whose anthropological interests were either avocational or subordinated to their primary scientific affiliations. The resulting interdisciplinary mélange could only have worked in this fluid proto-professional period. A scholar still could contribute to anthropology without being, in a sense recognizable today, a professional anthropologist.

The second cluster of institutional invention in late-nineteenth-century anthropology arose around Putnam at Harvard's Peabody Museum of Archaeology and Ethnology. A natural scientist by training and an entrepreneur by inclination, Putnam was an effective organizational leader whose fine hand orchestrated early anthropological developments at the Chicago World's Fair in 1893 and at the University of California, Berkeley, as well as in Cambridge. Although Putnam's own work was in archaeology and physical anthropology, which became the specialization of the graduate program at Harvard, his definition of anthropology remained a broader one. Ethnology and archaeology students (for example, John Swanton and Roland Dixon) were sent to work with Boas at Columbia, although their degrees came from Harvard. Thus, although Putnam did not explicitly enter into debates about the scope or the terminology of the emerging professional discipline, he was sympathetic to its broader aspirations. Harvard archaeology exemplified and reinforced the centrality of archaeology to American anthropology.

The whole question of scope and professionalization reached its most dramatic enactment in the ongoing clash between Powell (and his minion WJ McGee) and a now nearly forgotten amateur anthropologist, Daniel Garrison Brinton of Philadelphia (Darnell 1988). At first glance, Brinton was Powell's antithesis in every possible way. His health precluded active fieldwork (after a brief youthful foray into Florida archaeology). Powell, despite the loss of an arm during the Civil War, was a rugged adventurer remembered for white-water rafting on the Colorado River. Brinton was an armchair scholar, a non-practicing physician and medical publisher before he turned to ethnology and linguistics. Powell thrived on the collaborative energy of the bureau and its myriad projects, as well as the political challenge of securing annual congressional appropriations adequate for the ongoing work of the bureau.

Brinton was intrigued by the civilizations and written literatures of Meso-America. Powell, by institutional constraint as well as by personal inclination, restricted himself to the anthropology possible within the political boundaries of the United States; he was a political creature who had his own version of American Manifest Destiny. Brinton pursued his solitary scholarship and reported it to various learned societies, with the American Philosophical Society of Philadelphia as the bedrock institutional affiliation of his career. Brinton's stature among the intellectual elite of North American scholarly institutions considerably exceeded that of the upstart Powell, arguably tainted by political machinations and adventurous expeditions purporting to be science. Brinton was a theoretician, Powell a pragmatic collector and synthesizer of facts. Brinton had a reputation for being cranky and self-absorbed whereas Powell was flamboyant and energetic.

Nonetheless, Brinton and Powell were of the same generation, both having served in the Civil War, and had some things in common. Both exceeded and challenged the conventional boundaries of what was coming to be called anthropology. Today we would characterize their audiences as interdisciplinary and only minimally professional. Powell's several biographers (Darrah 1951, Stegner 1954, Terrell 1969, Worster 2000) memorialize his geological rather than anthropological contributions. Both wrote for scientifically inclined readers who were not, or not primarily, anthropologists. Both, after an early fascination with archaeology, turned to ethnology and linguistics but found themselves drawn into the professionalization of archaeology (with its consequent inexorable inclusion among the anthropological sciences) by public enthusiasm for archaeology—Brinton by the institutional vagaries of Philadelphia amateur science and Powell by mandate of Congress.

Despite the absence of many of the distinctive features of professional science in his career, Brinton successfully juggled publication outlets, public lecture fora, and an extensive national and international network maintained through correspondence. Moreover, he defined himself, far beyond the city limits of Philadelphia, as the spokesperson for the nascent professional discipline of anthropology. In collecting his essays for publication in 1890, Brinton's chosen label was "Americanist." Gradually, anthropology became his cover term for broad interests in ethnology, racial typology, and linguistics. In retrospect, this position anticipated the three classificatory variables for the broad anthropological science now associated with Boas (1896, 1911) and his methodological critique of evolutionary comparative method. Boas's argument for the autonomy of race, language, and culture meant that the new anthropology could not exclude any of these variables without descending into futile reductionism. Archaeology, in this formulation, added time depth to the anthropological study of race, language, and culture.

Despite brief forays into the field and archaeological topics, Brinton was not, by contemporary definition, an archaeologist. Because he was dedicated to defining anthropology as the "science of man" in the broadest sense, archaeology fell within his self-proclaimed mandate for the discipline. He saw archaeology as a method rather than a substantive branch of anthropology. It absorbed the slack between written history as understood in traditional European historical philological scholarship and ethnology, increasingly based on firsthand fieldwork. Conceptually, he argued, anthropology was of a single piece, with its theoretical integrity spanning the subdisciplines despite considerable confusion over the course of his career in defining and interrelating them. Because he was not a fieldworker, in archaeology or ethnology, Brinton focused primarily on questions about the mental products of human culture, what the social evolutionary theory of the day called the "psychic unity of mankind." As for Powell, the potential racist implications of his evolutionary thinking were subsumed under the Americanist orientation. Both took for granted that "the American Indian" resided, in evolutionary terms, at a single stage of progress toward civilization, being, for example, far superior to African Negroes and their degenerated American descendants (cf. Baker 1998). At the end of his career, Brinton (1902) updated his technical vocabulary for the cross-cultural science of mind; his posthumous book is entitled *The Basis of Social Relations: A Study in Ethnic Psychology.*

The public appeal of archaeology greatly outstripped that of other aspects of anthropology. Brinton took advantage of the archaeological enthusiasm of the University of Pennsylvania's dynamic provost William Pepper and his philanthropist crony in the founding of the University Museum, Sara Stevenson. In terms of the scope of the field he had preached throughout his career, Brinton felt able to claim archaeological expertise, although it emanated from his Philadelphia armchair. Owing to the institutional machinations of Pepper and Stevenson to create a great anthropological museum in Philadelphia, Brinton became the first university professor of anthropology in North America in 1886, a year before Putnam's appointment at Harvard. Brinton's academic position was entirely honorary, with no evidence that he ever offered courses or provided professional training. The University Museum focused its archaeological aspirations in more aesthetically prominent domains: Egypt, the Near East, and China (all still represented in the ethnographic displays at the museum). Putnam, in contrast, commanded institutional resources for teaching in combination with museum research that were not available to Brinton, whose position remained marginal to the grand schemes in Philadelphia. Although Harvard increasingly specialized in Meso-American archaeology, American Indian work always found a place there.

For the history of archaeology, the most interesting thing about Brinton's nominal professorship is its title—Professor of Archaeology and Linguistics. The linguistic side of this appointment was congruent with Brinton's publications and long-term Meso-American specialization. In the University Calendar, he listed potential courses in Nahuatl and Mayan language, literature, and hieroglyphics. These were all familiar subjects for him. Brinton was among the first to identify significant phonetic elements in Mayan writing and had published a *Library of Aboriginal American Literature* in the 1880s, with five of the eight volumes presenting his own Meso-American textual work; unfortunately, this work was highly technical and did not lend itself to museum display.

Indeed, Brinton's linguistic classification, published in direct competition with Powell's in 1891, included both North and South American languages, even though much less was known about the latter. It is, then, perhaps not surprising that Brinton's ambitious academic offerings also included the classification and general structure of the languages of North and South America. His view that American languages were polysynthetic, reflecting their evolutionary similarity, predisposed him to think in continental terms, with minor geographic variants. One calendar course proposal even included Kechua and Algonquian (in practice, Lenape Delaware, spoken near Philadelphia, with which he had at least a nodding acquaintance).

On the archaeological side, Brinton proposed a course in methodology and a survey of American archaeological explorations, "the characteristic remains in the leading archaeological provinces of North and South America" (cited in Darnell 1988:54). He further outlined a course to address the relations between archaeology and ethnology. The latter ostensibly was not part of the professorship but reflected his broad definition of the appropriate scope of anthropology. Later calendar copy, presumably reflecting local interests, retreated to North America, even singling out the "antiquities of the Eastern United States" (cited in Darnell 1988:55). These local materials constituted the bulk of the University Museum collections that fell under his purview.

Despite Brinton's efforts, however, and presumably much to his dismay, colleagues in Near Eastern, Egyptian, and Classical archaeology accrued more prestige both in terms of the organization of the museum and the introduction of their subjects within the university. The museum's policies for antiquities collection and display entailed firm geographic distinctions and specializations that were uncongenial to Brinton. In addition to his Americanist scholarship, he was active in the American Oriental Society and doubtless felt comfortable applying the psychic unity paradigm to Old World cultures foregrounded by the museum. Parceling out of ethnographic areas ran counter to the kind of generalizations inherent in Brinton's anthropology.

In "Anthropology as a Science and as a Branch of University Education in the United States," Brinton (1892a) attempted to establish the systematic study of anthropology at the University of Pennsylvania, simultaneously offering his prospectus as a model at the national level. He did not mention that anthropology already had been taught for some years at Harvard, although with a primary focus on archaeology. Brinton's scheme was abortive, proposed before its time had come. Institutional resources were not available in Philadelphia, and Brinton's nominal professorship became more rather than less peripheral as time went on. Nonetheless, his scope and accompanying terminology were close to those eventually adopted by apparent consensus in American anthropology. Yet, the history of Americanist anthropology has failed to acknowledge Brinton's theoretical precedence. Credit for ideas depends strongly on institutional backing for their implementation; both Powell and Putnam had far more backing than Brinton. That the latter's vision was not shared locally does not, however, detract from the prescience of his analysis, in the final years of his own career, of the need of Americanist anthropology for an academic base.

Brinton understood that the constellation of anthropological institutions in and around Philadelphia offered the potential for leadership in the development of anthropology as a science. He was fully aware that professional training in universities would be necessary for the next generation but was unable to mobilize the local university to create the wave of the future he foresaw. Because the anthropological museum in Philadelphia was affiliated with the university, Brinton's potential institutional tools to increase the viability of anthropology at the University of Pennsylvania were amateur societies with considerable local prestige but decreasing salience for the actual production of science. He deployed his connections at the American Philosophical Society and the Academy of Natural Sciences of Philadelphia but failed to parlay them into vehicles of national or professional stature that would meet new emerging standards.

Brinton's reputation was further eclipsed by Boas, whose presentist description of the history of anthropology in *Science* in 1904 deliberately effaced prior Americanist work under an evolutionary paradigm, dismissively critiquing Brinton's position without even naming him. The future would begin with Boas himself. Since that future, at least in sociocultural anthropology and linguistics, according to the division of labor between Columbia and Harvard, was increasingly dominated in the first two decades of the twentieth century by Boas's students, his reading effectively rewrote disciplinary history. (See Cole 1999 for the vicissitudes of Boas's slow move toward such personal security and professional reproductive success. Only in retrospect did the process seem inevitable or straightforward.)

Within the preprofessional paradigm in which Brinton's reputation was initially established, his professorship in ethnology and archaeology at the Academy of Natural Sciences of Philadelphia in 1884 was at least as significant as the nominal appointment at the University of Pennsylvania. The academy had a substantial faculty, all honorary lecturers to a general public audience interested in science. Well-to-do Philadelphia matrons filled the lecture rooms; popular science was a prestigious leisure activity.

Archaeology was more of an attraction to a general audience than the ethnological and linguistic core of Brinton's personal scholarship. Nonetheless, he applied for the positions, probably created with him in mind, and used the lectures as formats for his books *Races and Peoples: Lectures on the Science of Ethnography* in 1890 and *The American Race: A Linguistic Classification and Ethnographic Description of the Native Tribes of North and South America* in 1891, dealing respectively with the Old and New Worlds (Brinton 1890b, 1891).

Brinton's influence on the scope of the anthropology we know today is perhaps best encapsulated in his terminological battles with the Washington anthropologists. The disputed terms themselves were in one sense almost extraneous to the struggle for control of the emerging profession. On behalf of Philadelphia, and employing the resources of his native city to back his position, Brinton claimed the status of a senior independent scholar to pronounce on the state of the discipline as a whole. With only minor modifications of his position, he returned repeatedly to the subject of subdisciplines and their appropriate labels. Although there is a certain amount of repetition in these efforts, Brinton clearly intended to carry his disciplinary message to as many different audiences as possible.

"Anthropology and Ethnology" in 1886 was directed to a local audience, in an encyclopedia published in Philadelphia. At this early point, Brinton restricted the term "anthropology" to the study of mankind as a biological species. In congruence with European usage, he (Brinton 1886:17–18) distinguished between ethnology as the study of human intellectual development resulting from the forms of "social relations" and ethnography as the description of particular "relations and customs." In *Races and Peoples* (Brinton 1890b), he divided ethnography into its physical and psychical elements as a prelude to cataloguing races and peoples [cultures]; theory and description proved inextricable.

"The Nomenclature and Teaching of Anthropology" (Brinton 1892c) appeared in 1892 in the *American Anthropologist* (old series, still the organ of the Anthropological Society of Washington; not until the establishment of the new series in 1898 could the journal be said to represent a national audience or consensus). By this point, Brinton had decided that the cover term "anthropology" "should" be employed "in the widest sense, defining several subdivisions within this broad field." Somatology included "experimental and practical" as well as developmental and comparative psychology, a scope more mentalistic than somatic by contemporary standards. Ethnology was broadly defined as both "historical and analytic." It encompassed "ethnic psychology," sociology, technology, the science of religion, linguistics, and folklore. Ethnography or "geographic and descriptive anthropology" remained separate. Archaeology was "prehistoric and reconstructive anthropology," presumably having both theoretical and descriptive components (Brinton 1892c:263–66).

"The Aims of Anthropology" (Brinton 1895) appeared in the *Proceedings of the American Association for the Advancement of Science* in 1895 and was also published in *Science, Popular Science Monthly,* and *Scientific American,* running the gamut of nonspecialized scientific publications. As retiring president, Brinton asserted unequivocally that the American Association for the Advancement of Science, the only national organization representing the full range of American science, shared his view. Like Putnam, he insisted that anthropology must include "the whole of man [*sic*], his psychical as well as his physical nature, and the products of all his activities, whether in the past or in the present." "We cannot but deplore," he wrote, the continental restriction of the term to physical anthropology. Brinton included history within anthropology. "Ethnology in its true sense represents the application of the principles of inductive philosophy to the products of man's faculties." It is "strictly a natural science, dealing with outward things—to wit, the expressions of man's psychical life . . . the natural science of social life." Ethnography dealt with subspecies and smaller groups (Brinton 1895:241–52). In principle, then, Brinton's typology did not restrict anthropology to the study of the so-called primitive.

The only local conglomerates of anthropological institutions capable of challenging Brinton's reliance on Philadelphia hegemony in speaking for the national discipline were in Cambridge and in Washington, D.C. Putnam did not engage in the debate on scope and terminology. Despite the institutional power mustered by Powell's bureau, firmly lodged within the Smithsonian and allied to the National Museum, Brinton and Philadelphia held their own on Powell's home turf, before the Anthropological Society of Washington. Absence of institutional backing failed to preserve the bureau terminology because its involuted neologisms failed to achieve widespread consensus. Brinton's schema much more accurately reflected the emerging consensus.

This meeting of opposing forces, however, has been simplified in professional memory, partially because the dispute between Brinton and Powell over terminology was reprinted in *Selected Papers from the American Anthropologist, 1888–1920,* edited by Frederica de Laguna in 1960. However dramatic the immediate confrontation, there were other players and positions, in Washington as well as by Brinton in different contexts. Although the *Anthropological Society of Washington Proceedings* record that Powell intervened in Brinton's paper to present his own version, Brinton's stature was sufficient to secure his invitation to speak authoritatively on a contentious topic.

Powell was not the originator of late-nineteenth-century terminological fetishism (quotations are taken from Dieserud 1908). Otis T. Mason of the United States National Museum entered into the classificatory game early. His Saturday lectures to the Anthropological Society of Washington in 1882 (Mason 1882) were reprinted in the *Smithsonian Annual Report* for that year. He restricted the term "anthropology" to the origin of man, also noting archaeology, "anthropobiology," ethnography (races) and ethnology ("the corresponding deductive science"), psychology (mind and brain), glossology (linguistics or philology), technology, sociology, comparative mythology, and hexiology or the "relations of the physical universe and social environment to human history, migrations, etc."

In the *Smithsonian Annual Report for 1881,* published in 1883, Mason (1883) elaborated his scheme with only minor variants. His neologisms distinguished description, classification, and prediction as the progressive stages of scientific method.

Having developed this systematic anthropological terminology, Mason wanted to know what other anthropologists thought about it. He even suggested proudly that a fourth column might emerge, of anthropogeny, and so forth, to clarify the origins of phenomena in the various categories. Needless to say, the impact of Mason's scheme was virtually nil. Only in the distinction between ethnography and ethnology was he in accord with the emerging

TABLE 2.1.
THE SCIENCE OF ANTHROPOLOGY

Observing and Descriptive stage	Inductive and Classifying Stage	Deductive and Predictive Stage
Anthropography	Anthropology	Anthroponomy
Archaeography	Archaeology	Archaeonomy
Psychography or Phrenography	Psychology or Phrenology	Psychonomy or Phrenonomy
Ethnography	Ethnology	Ethnonomy
Glossography	Glossology	Glossonomy
Technography	Technography	Technography
Sociography	Sociography	Sociography
Pneumatography or Daimonography or Mythography	Pneumatography or Daimonography or Mythography	Pneumatography or Daimonography or Mythography
Hexiography	Hexiography	Hexionomy

consensus. His neologisms are today merely bizarre. Nonetheless, the subdisciplines, with the partial exception of physical anthropology, are present, albeit obscured by the mélange of fields no longer considered separate from the umbrella category of sociocultural anthropology. Familiar terms occur mostly in the center column, suggesting that Mason and his contemporaries thought anthropology was stuck at the inductive level, effective at classifying but unable to attain the deductive and predictive elegance of the natural sciences.

By the time Mason (1884) published "The Scope and Value of Anthropological Studies" in the *Proceedings of the American Association for the Advancement of Science* in 1884, reprinted in *Science*, he had soft-pedaled his classificatory elaboration. Anthropology was now "the natural history of the genus homo," in which many fields "enter into the scope of anthropology, whose task it is to comprehend the work of all in its synthesis." The new science claimed vast interdisciplinary territory just as the disciplines were beginning to emerge in their professional forms.

Mason's (1891) final effort in the *Smithsonian Annual Report for 1890* returned to his terminological obsession as a way of claiming broad disciplinary ground. But this time he switched the grounds of definition, distinguishing "Structural Anthropology" or "what man is" from "Functional Anthropology" or "what man does." The former included onogeny, anatomy, physiology, anthropometry, psycho-physics, and ethnology ("the natural divisions of mankind"), with the latter incorporating glossology, technology, aesthetics, science and philosophy, sociology, and religion. Here, Mason clarified what he meant by history in the context of the scope of anthropology: archaeology, paleography (the decipherment of inscriptions), folklore, and history covered "the past of human life and actions," again not distinguishing between description and theory in archaeology. Mason claimed for anthropology only that part of psychology that measures the activities of mind.

Powell himself did not enter into debate with Mason, although he wrote on "The Humanities" in the popular outlet *The Forum* (Powell 1890) that, in his view, ethnology "as an attempt at a classification of mankind by races . . . has failed to become a science." Rather, "the processes of human evolution" had proved to be "not biotic but cultural." In this new context, anthropology as the broad study of man gained the potential to define "a new realm of science." In fact, Powell asserted, "it might well be called the science of culture and perhaps still better the science of the humanities."

Powell turned to terminological involution only in his 1892 debate with Brinton in the *American Anthropologist.* Somatology (physical anthropology) and archaeology are clearly recognizable with their present scope. Psychology stands alone, but the other aspects of what would now be called cultural anthropology (technology, sociology, philology, literature, esthetology, natural religion, and sophiology or "the evolution of thought as exhibited in the lore of mankind") are grouped under ethnology, which is still distinguished from ethnography. Linguistics or philology is not separated from ethnology.

Writing about anthropology in the *Universal Cyclopaedia,* Powell (1900) clarified what he meant by psychology; it included the study of the human soul, in comparison to the lower animals, and was simultaneously allied to philosophy and natural science. The mentalism implicit in studying humanity, therefore, led Powell to separate human capacity for spirituality from the merely biological, which was the proper realm of the natural sciences. The latter presumably fell under somatology. This kind of distinction would fail to meet emerging criteria for an anthropological science bridging mind and body.

Powell (1900:1) continued to define ethnology according to the physical traits of "races or groups." "The tendency at the present time is to apply the word to that science which treats of the culture of the tribes and nations of mankind, or the origin, development and condition of industrial and decorative arts and the arts of amusement; the arts of society; the arts of language; the arts of literature; the fine arts; natural religions; and, finally, the opinions of mankind." Archaeology presumably included the study of any of these areas from the standpoint of the past.

The cudgel for Washington anthropology was taken up increasingly by WJ McGee as Powell's health declined during the final years before his death in 1902. McGee's (1897) vice-presidential address to the American Association for the Advancement of Science on "the Science of Humanity" was published in *Science.* His anthropology had three major divisions: "physical anthropology," psychology, and "the science of humanity" (including the more conventional of the earlier components of ethnology—esthetology, technology, sociology, philology, and sophiology). Despite the talk of humanity, McGee insisted that all three areas treated man [*sic*] as an organism. In fact, he identified no consensus about whether anthropology could expect to discover anything beyond the organism, although psychology might arrive at such generalizations. Seemingly, McGee was anxious about the scientific status of anthropological investigation of mental phenomena; apparently, Powell's idealism/mentalism had gone too far.

Later, McGee (1905) returned again to "Anthropology and Its Larger Problems." He was prescient about changes in the discipline, mostly attributable to the growing stature of Boas. McGee was close to these developments during the years when Boas was the bureau's honorary philologist with McGee as his contact. The two also worked together around 1898 to establish the new series of the *American Anthropologist.* McGee refused to abandon the long-established Washington neologisms, despite their failure to capture wider acceptance. Somatology was equated with European physical anthropology (without acknowledgment that the adjective "physical" was not used in Europe). Psychology was carefully defined to exclude everything except "the strictly inductive mind science of current American schools" (presumably behaviorism rather than mentalism, again an insistence on the scientific aspirations of Washington anthropology, on the model of the natural sciences). Ethnology retained its "subsciences" of esthetology, technology, sociology, philology, and sophiology. McGee was particularly fond of pentads. The cover term was not anthropology but "demononomy, or principles of peoples." Both ethnology and somatology had "prehistoric aspects," which included archaeology. McGee's overall characterization of "the field of the fin-de-siécle anthropology" attempted to move beyond outdated racial classification but still retained the evolutionary notion of progress: in "the modern anthropology, sometimes styled the new ethnology, the peoples of the world are not divided into races . . . but grouped in culture-grades [which] correspond with the great stages of human progress."

On the one hand, McGee clung to the theoretical foundations of Powell's early bureau, based on Lewis Henry Morgan's *Ancient Society* (1877). At the same time, however, he moved beyond the traditional American Indian mandate of the bureau to claim a more general scope for anthropology as the study of all human groups. Classificatory terminology, then, provided a convenient means to maintain and perhaps even extended the hegemony of Washington within the professional discipline of anthropology now well established, albeit recently, at the national level. The founding of the American Anthropological Association in 1902 properly marks the conclusion of this jockeying for local status in the movement toward professionalization, which is perhaps best exemplified by Brinton's Philadelphia-based efforts toward national and professional maturity for anthropology. A few years earlier, the new series of the *American Anthropologist* could be established only by balancing Washington dominance against Boas's nascent and still largely illusory power base in New York. The rest of the editorial board represented other cities with a stake in the control and evolution of emerging national institutions. Philadelphia, unsurprisingly, was represented by Daniel Brinton. After Brinton's death in 1899, he was succeeded on the board by his protégé and nominal successor at the University Museum, Stewart Culin. The geographic logic underlying the tenuous national identity of the journal simultaneously confirms the important role of Philadelphia as mediated to other local power blocs by Brinton and documents the establishment of a solid power base for a national professional organization transcending those same local boundaries.

3

Unsung Visionary

Sara Yorke Stevenson and the Development of Archaeology in Philadelphia

ELIN C. DANIEN AND ELEANOR M. KING

IN THE LAST YEARS OF the nineteenth century, the modern university was still being created. Vestiges of past academies lived on in the reconstruction that led to the new. The modern museum and archaeology as a profession were in their infancy. Archaeology, still centered in museum collections, had not yet committed itself to the academic venue it inhabits today. The decades from 1880 to 1920 have been called anthropology's "museum age" (Dockstader 1967, Sturtevant 1969). Indeed, many universities looked somewhat askance at, and hesitated to include, this new discipline in their offerings. Harvard University, home of the Peabody Museum, the nation's first museum specifically devoted to archaeology and ethnology, established a professional chair in anthropology only in 1887, twenty-one years after the museum's founding (Hinsley 1985, 1992). During these formative "museum" years, the independent scholar, the museum archaeologist, and the university professor each functioned in a crucial and defining moment that would change them forever—or at least for the next century (Dockstader 1967, Osgood 1979:12).

One of those who played a major role in making archaeology an important segment of Philadelphia's cultural universe was Sara Yorke Stevenson, a woman whose intellectual contributions and breadth of interests would have been remarkable at any time (figure 3.1). She was an Egyptologist, a spokeswoman for archaeology who helped make financial support for archaeological expeditions popular among Philadelphia's elite, a founder of the University of Pennsylvania Museum of Archaeology and Anthropology (originally known

3.1. Sara Yorke Stevenson. An undated photograph, probably dating to the 1890s. University of Pennsylvania Museum. (Neg. #S4-133978)

as the Free Museum of Science and Art) and the driving force for the construction of its building, and one of the first to view museology as a profession. Her contributions are all the more exceptional because she was a dominant figure in Philadelphia's cultural life during a period when women were limited to a very small piece of the public intellectual pie.

Early Years

Born in Paris in 1847 to Edward and Sarah (Hanna) Yorke, Sara Letitia Yorke spent her childhood in France among antiquarians and scholars who encouraged her early intellectual development. Her determination and self-possession were evident at an early age. In 1857 her family returned to the United States, where Sara quickly rebelled at what she considered to be an "inadequate" day school. The ten-year-old was able to convince her parents that she should return immediately to Paris and a boarding school that would provide her with the intellectual life she craved. The home of her Parisian guardians, M. and Mme. Achille Jubinal, was a center for those who had distinguished themselves in literature, art, science, and politics. M. Jubinal, an authority on ancient tapestries, arms, and armor, maintained a private museum and first fostered Sara's interest in archaeology, and especially Egyptology (Wister 1922:9). These peaceful years ended suddenly on March 4, 1862, when Sara's brother Ogden was murdered by brigands in Mexico, where he was assisting his father in a project to build a trans-isthmian railroad. At the age of fifteen, accompanied by a chaperone (the two being the only women on board), Sara made the long and somewhat perilous sea voyage from France to Mexico to join her parents. Her adventures during the next five years in Mexico, living through the tumultuous era of the French intervention, undoubtedly contributed to her growing self-reliance and resourcefulness. Her tact, self-control, tenacity in the face of obstacles, and ability to convince others of the rightness of her actions were honed during her Mexican sojourn, when skillful diplomacy was imperative to deflect the constant physical dangers (Stevenson 1899).

In 1868 she came to Philadelphia to live with three aged and rather eccentric relatives. Her lifetime habit of late hours was formed then, when she would share a late supper with her uncle, blind Admiral Yorke, and swap stories with him until long past midnight (Wister 1922:10). She was twenty-one, a young woman of impeccable social credentials, excellent education, a fine mind, and a strong character, but little money. In moving to Philadelphia, one of her aims, certainly, was to make a good marriage, and that she did.

Within two years she met and married Cornelius Stevenson, a Civil War veteran, lawyer, and member of a prominent and wealthy old Philadelphia family whose antiquarian interests echoed her own. Although she was very much a product of her social milieu, and knew how to operate within and manipulate its boundaries, it is clear that she never lost a sense of alienation in the rigid society of Victorian Philadelphia. Many years later, she became a literary editor for the Philadelphia *Public Ledger.* For her weekly social column, she adopted the pen name of Peggy Shippen, an eighteenth-century

Philadelphia belle who later became the wife of Benedict Arnold (Stevenson 1907–21).

Intellectual Life

In the decade following her marriage, at the same time that she excelled in all the acceptable activities of Philadelphia matrons—balls, dinners, entertainments, charity work—"much as society people do today, but with more decorum and better manners" (Wister 1922:10), she discovered and was welcomed by Philadelphia's small group of intellectual elite. The relationships she established then served her well in her career as museum builder; they were friendships that endured throughout her long life.

Of primary importance was her acceptance as a member of the Mitchell-Furness coterie, a group of physicians, scientists, writers, scholars, and anthropologists who defined the intellectual, social, and civic life of Philadelphia from the 1870s to World War I. Both intellectual accomplishment and social position were among the criteria for inclusion in the group, whose numbers never exceeded eighteen. Stevenson engaged in discussions with members, such as Shakespeare scholar H. H. Furness; Owen Wister, author of *The Virginian;* noted physician and poet S. Weir Mitchell; and Talcott Williams, first dean of Columbia University's School of Journalism.

The coterie was unusual in its acceptance of accomplished women as intellectual equals. Several women were members, among them Agnes Irwin, first dean of Radcliffe College and co-editor of an early effort at creating a women's history, and the noted American essayist Agnes Repplier. Sara Yorke Stevenson's charm, intelligence, and expert knowledge of archaeology and Egyptology were recognized and nurtured in this atmosphere. As a member of the socially powerful coterie, she was allowed greater social and intellectual freedom than women ordinarily enjoyed in Victorian Philadelphia (Van Ness 1985, 1988).

Her contributions to Philadelphia's cultural life began early and took on a momentum that ceased only with her death. In the final decades of the nineteenth century, the field of anthropology had not yet become the domain of academe. The many anthropology clubs and professional societies, with their lectures, discussions, and research, provided opportunities for the amateur and independent scholar. Like her colleague Daniel Brinton and other anthropologists, Stevenson had received a classical education along traditional lines. She was one of the generation of "armchair archaeologists" who never carried out fieldwork (Van Ness 1988:345). Instead, her research, like Brinton's, was based largely on the analysis of material collected or excavated by others. Her interest in Egyptology was stimulated by F. W. Putnam, who provided encouragement and urged her on to further research (e.g., Stevenson 1892a, 1892b, 1893, 1894, 1896; Stevenson, Jastrow, and Justi 1905). She presented her thinking on Egyptian topics in a course offered at the University of Pennsylvania and in other lectures that, according to a contemporary, "made a sensation" (Starr 1892).

Stevenson's reputation in the United States soon came to rival Amelia Edwards's in England, and honors quickly accrued to "our only lady Egyptologist" (Starr 1892). In 1884 she was admitted to membership in the American Association for the Advancement of Science, and she became a fellow in 1895. In 1891 the Oriental Club of Philadelphia, a group of some thirty professors from leading universities, changed its bylaws stipulating male membership in order to admit her. Two years later, in 1893, a special act of Congress was required to allow her, as a woman, to serve on the Jury of Awards for Ethnology at the World's Columbian Exposition in Chicago. Her colleagues on the jury elected her vice-president (Stevenson 1901). In 1894 she was the first woman to lecture at Harvard's Peabody Museum when Putnam invited her to speak on "Egypt at the Dawn of History" in the anthropology lecture series (Jordan 1911:1615, Van Ness 1988, Wister 1922). In 1894 she became the first woman to receive an honorary doctorate from the University of Pennsylvania. In 1895 she was the second woman invited to become a member of the American Philosophical Society (Meyerson and Winegrad 1978:117–29).

Stevenson always reached out beyond the scholarly community and seized every opportunity to promote anthropology and archaeology to an educated public. She was active in a number of organizations, including the Numismatic and Antiquarian Society and the Oriental Society, and helped establish many others. In 1886 she was one of the founders of the Contemporary Club, which invited scholars to speak to a lay audience on a broad range of topics, many of them dealing with anthropology. In 1888 she was one of the founders of the University Archaeological Association, a forerunner of the University Museum. She joined Daniel Brinton, Carl Lumholtz, Stewart Culin, and others in 1889 forming the American Folklore Society. She was also a founding member of the Pennsylvania branch of the Archaeological Institute of America (Jordan 1911, Wister 1922, Van Ness 1988).

Stevenson and the University

Stevenson's most lasting contribution, however, was the establishment of the University Museum of the University of Pennsylvania. Stevenson's involvement with the development of the museum began even before the thought of a separate museum was contemplated. In 1887 a group of Philadelphians, including many of the members of the Mitchell-Furness Coterie, proposed to underwrite an exploring expedition to Babylon. In return, all finds that could be exported would be brought to Philadelphia and become the property of the University of Pennsylvania, provided a suitable place was found to house them (Madeira 1964:15–16, Kuklick 1996:27).

One of those involved in the discussions was William Pepper, then provost of the university and a member of the coterie. His ambitions for the university had resulted in thirteen new departments, twenty new buildings, and a much broader curriculum. In planning a museum, however, his vision was far surpassed by Stevenson's. Longtime acquaintances and colleagues, they became close friends—partners, allies, and coconspirators—in the long and often difficult journey to their goal.

Pepper had thought to display the artifacts brought back by the Babylonian Exploration Fund on one or two floors of the new university library building. The rapid accumulation of material from the continuing expeditions soon made it clear that more space would be needed. Pepper suggested only the addition of an archaeological wing for the library, but as letters between Pepper and Stevenson indicate (University Museum Archives [UPM] Director's Files, Pepper-Stevenson Correspondence), Stevenson's concept was much more sweeping. She envisioned a completely separate building where a growing collection could be displayed. And unlike natural history museums where archaeology was only part of the overall display, this museum would have an exclusive focus on archaeology and ethnology (Conn 1998:86). She also conceived its research to be global in scope, with collections drawn from every culture in both Old and New Worlds (UPM, Director's Files, Box 1).

In 1889, while the first museum expedition was still digging at the ancient Mesopotamian site of Nippur, the university created a quasi-official Department of Archaeology and Paleontology. Despite the title, the department was not under Penn's academic umbrella but rather functioned as an independent museum, albeit still without walls. The university provided the services of certain professors, notably Herman Hilprecht, to act as curators. Their salaries were covered by their faculty appointments in their respective academic departments. Stevenson was the volunteer curator of Egyptology and the Mediterranean collections. The only curator paid directly by the department was C. C. Abbott, half of whose annual salary of one thousand dollars was provided by private subscription.

The idea of paid professional staff was almost anathema in a circle that reserved cultural life for those who could afford it. It is thus all the more remarkable that in 1893, Stevenson, who was part of this society and subscribed to its mores, championed the cause of a young German Jewish scholar. Franz Boas, following his work at the Columbian Exposition, was without an institutional home. At Frederic Ward Putnam's suggestion, Stevenson put Boas's name forward to head the newly created Wistar Institute. Despite months of negotiating, she did not prevail, the university having decided his salary was an unnecessary expense. Thus Penn lost Boas to the American Museum of Natural History and Columbia University (Hinsley and Holm 1976, Hinsley, this volume, UPM Director's Files, Box 1).

The museum's governing structure reflected the fractured nature of its sponsorship. The university trustees were reluctant to incur costs, and so, in order to raise money for expeditions and collections, in 1888 the University Archaeological Association was incorporated as the museum's fund-raising arm. Its incorporation papers confirm the need to cooperate with the university but firmly maintain its separate identity. The university was to appoint only twelve of the thirty-six members of the museum's board of managers; control would rest with the twenty-four members chosen by the association (Madeira 1964:19). Despite numerous bylaw changes over the years, the majority of the museum's board continues to be drawn from outside the university.

Stevenson immediately set about creating the museum and acquiring the collections to fill its future galleries. She corresponded with Flinders Petrie and arranged for certain of his objects to be sent to Philadelphia in return for financial support for his Egyptian excavations, and in 1890 he sent the museum its first Egyptian antiquities. She fostered cooperation with other museums by inviting them to join with the University Museum in this endeavor and to obtain a share of the finds. In 1897 and 1898 she traveled to Rome and then to Egypt on behalf of the museum. Her tart comments on the cultural maze she encountered are unambiguous: "Never in my life saw such intriguing. Politics-science-personal rancor is all mixed up." On another occasion she wrote, "One must have a genius for diplomacy to steer an enterprise through Egyptian conditions today."

Although her diplomatic genius was insufficient to arrange an excavation, she was able to acquire a large number of artifacts. When she returned to Philadelphia she brought back scores of packing cases filled with objects for the collections (O'Connor and Silverman 1979). Under her aegis, the museum sponsored fieldwork in many lands and acquired collections from the pre-Columbian cultures of Peru, Yucatan, and Florida as well as from Crete and Russia (Madeira 1964, Winegrad 1993:37–41). She organized lectures and promoted the value of archaeology and the importance of the university museum among her friends, the public, and the government.

Creating a Museum Building

The machinations required to build a home for the museum called upon all of Stevenson's patience, tenacity, and diplomacy. During the better part of the nineties, she and Pepper skirted the minefields of political and academic intrigue in order to acquire from the city the large tract of land that had been the Blockley Township Almshouse and then to guard it against the insatiable land needs of the growing university (Madeira 1964:22–24, Winegrad 1993: 21). Pepper was keenly aware of the problem and warned Stevenson that "the very existence of the institution depends on its political independence" (Pepper to Stevenson, 1893 [no month or day] UPM Director's Files, Box 1).

Over the years, the balance of power between university and museum was constantly in flux, a state that in many ways continues to this day. Funds for the proposed building were elusive. After months of negotiations, Stevenson went to Harrisburg and convinced the state to donate $150,000 toward the museum building. There followed a test of wills, as the association vied with the university to control these funds. The association lost, and the university was able to dictate the choice of architects and contractors. But despite this setback, and the loss of some of its autonomy, the association continued its fund-raising. Together, Pepper and Stevenson were an unstoppable juggernaut, a fact they well recognized. In one of the many daily notes that flew between them, Pepper exulted, "No one in the company in which we are can stand against us" (Pepper to Stevenson, March 1894, UPM Director's Files, Box 1). Using their positions in society and their many intellectual friends, they succeeded in raising an additional $200,000 to complete the first-phase building, despite fiscal vicissitudes that included the financial panic of 1893.

In 1899, one year after Pepper's death, the officially titled Free Museum of Science and Art opened its doors. The building, only one third of the majestic tripartite design by architect Wilson Eyre Jr., was a superb example of monumental architecture. One entered a large garden through an Oriental gate, walked past a rectangular lily pond, and came to a wide double stair leading up to a landing and a pair of outsize bronze doors. But behind those doors, trouble was still brewing in the hybrid structure of the institution. The year 1899 also marked the merger of the University Archaeological Association with the university's Department of Archaeology and Paleontology. Ironically, the association's supporters, not the university, still provided the funds for expeditions (Winegrad 1993:20–21). In 1905 the tension between university and museum came to a head with the notorious Hilprecht controversy.

The Hilprecht Controversy

Herman V. Hilprecht was celebrated as the excavator of Nippur. As professor of Assyriology at the University of Pennsylvania and curator of the Babylonian collections at the museum, his self-promotion, suspicion, and jealousy created a strained atmosphere. He antagonized his colleagues, both inspired and humiliated students, and engaged in activities that led to hints of plagiarism and misrepresentation. There were suggestions that his trumpeted great find of a "Temple Library" at Nippur was vastly exaggerated and unsupported by the data. When one of his colleagues accused him of having appropriated some artifacts originally meant for the museum, opposing forces were quick to form. The museum's relation with the university was a second, unspoken consideration in the forthcoming battle.

University provost Charles C. Harrison supported Hilprecht. The university conducted its own panel of inquiry and found no cause for action. Stevenson, as president of the museum's board of managers, together with the other members of the board, believed that the evidence supporting this accusation, coupled with earlier charges, was sufficient to warrant Hilprecht's dismissal. Stevenson had fiercely defended the museum's independence against the centralizing momentum of the university. All of the original incorporating papers emphasized that separation.

At the museum's inception the university had insisted on financial disassociation and that the museum's funds come from outside sources. But the university under Provost Harrison had been attempting to exert greater control over the museum, and this controversy provided an ideal opportunity. Penn maintained that it alone had control over the hiring and firing of curators, overruled the museum's dismissal, and reinstated Hilprecht. Stevenson was outraged and called a special meeting of the board of managers. On the principle of autonomy, despite her belief that the museum was the most important work of her life, she resigned. She was accompanied in her departure by a number of board members and curators, many of them members of the Mitchell-Furness Coterie, as well as more than one hundred museum members. From that point on, the university exercised ever greater control over the museum's activities (Madeira 1964:26–29, Kuklick 1996:123–41).

Beyond the University

Stevenson's scholarly contributions did not end with her self-imposed exile from the museum, however. She became curator at the Pennsylvania Museum, later known as the Philadelphia Museum of Art. And for the first time, she accepted a salary, a step made necessary by family financial problems. In 1908, at the art museum's affiliated School of Industrial Art, Stevenson offered the first museology course given in the United States. Her students visited every museum in Philadelphia, met with directors and curators, critiqued exhibits, identified problems of preservation and conservation, analyzed administrative methods, and learned the importance of museum cost efficiency (Wister 1922:37–38). "The time is not far off," Stevenson (1909) said, "when a standard of museum excellence will be attained, and when the great museums of this country, having settled on this standard, will undertake the training of assistants along a recognized line of theoretical and practical efficiency."

Sara Yorke Stevenson's enthusiasm for life and work continued in other areas as well. It was only now, in her sixtieth year, that she began the career for which she is most famous in Philadelphia. Her mornings were spent at the Pennsylvania Museum, but afternoons found her at her desk at one of Philadelphia's oldest newspapers, the *Public Ledger.* As literary editor and columnist, she supported women's suffrage, chided unseemly behavior, and reported on the economic and social issues of the day. She was a dynamic and long-lived presence in the lives of her readers, completing her last column less than a week before she died in 1921. At the same time she became more involved than ever in civic affairs, both locally and nationally. She also gained international recognition for her work for the French War Relief Committee, organizing relief efforts during World War I and receiving a coveted Legion of Honor medal from a grateful French government.

Conclusion

At her last birthday celebration, Stevenson was toasted by her good friend Agnes Repplier as "the woman who has been president of everything except the United States and the Women's Christian Temperance Movement," a position her French upbringing no doubt persuaded her to avoid. Indeed, the *Public Ledger* in its obituary stated that she occupied such "a unique position of admiration and esteem, of honor and of love, in Philadelphia society" that she was "a law unto herself—yet she never abused the power she indisputably wielded" (Wister 1922).

Sara Yorke Stevenson's remarkable life and career spanned a fascinating period in the burgeoning discipline of anthropological archaeology and in the evolution of Philadelphia's cultural life. Her early years, and indeed the founding of the museum, can be seen as growing out of amateur enthusiasm and independent scholarship, coupled with a newly defined need for order and organization—in other words, professionalization—of the discipline.

As the discipline's parameters became more established, the independent scholars were outnumbered and eventually replaced by the professionals, among them the first Ph.D.s in anthropology at the turn of the century (Collier and Tschopick 1954, Kroeber 1954, Dockstader 1967, Gruber 1967, Hallowell 1967). At the museum, George Byron Gordon, possessor of one of the early Harvard doctorates in anthropology, moved in to fill the power vacuum left in 1905 by the departure of Stevenson and several talented curators in the wake of the Hilprecht controversy. It was Gordon who became the first official director of the museum in 1910. He, in turn, hired William Farabee, another Harvard Ph.D., and many of Boas's students, most famously—or perhaps infamously—Frank Speck. It was the dispute between Speck and Gordon that destroyed any possibility of respectful coexistence between museum and anthropology at Penn. Speck formalized the rift by having the anthropology department physically moved out of the museum and into College Hall in a parade of people, papers, and paraphernalia that has become part of museum folklore. Gordon's directorship solidified the museum's position within the academic sphere of the university.

By the time of Stevenson's death in November 1921, academic credentials were essential in ethnology and archaeology alike as anthropology became solidly established in university departments throughout the country. Some museums merged with universities; others continued as autonomous educational entities, but few employed independent scholars as curators as they had in Stevenson's day. Nowhere was this more evident than in Benjamin Franklin's city, where the separation of anthropology from the museum devoted to it only made these changes more pronounced, leaving the cultural life of the city forever transformed.

4

In the Heat of Controversy

C. C. Abbott, the American Paleolithic, and the University Museum, 1889–1893

David J. Meltzer

In 1872 Charles C. Abbott (1843–1919)—a naturalist and collector in Trenton, New Jersey—took the first tentative steps toward establishing an American Paleolithic. Over the next ten years his archaeological collecting and writing intensified, and he laid the foundation for the American Paleolithic in a string of papers published through the 1870s and into the 1880s. There Abbott established the structure and much of the tone for what he believed about the archaeology of the Delaware valley, and would believe until his death nearly fifty years later. In those early papers, Abbott showed that the objects he was collecting deep in the gravel deposits of the Delaware River valley were unmistakably the result of human manufacture. That these apparently primitive tools were morphologically unlike artifacts in use by contemporary Native Americans, and since they were only rarely found in burials or sites alongside any "Indian" artifacts, they were obviously not the debris of ancestral Native Americans either. Instead, he claimed, the artifacts were the work of an unrelated group. Judging by how similar those artifacts were to primitive European Paleoliths, and how deep they were found in apparent glacial age deposits, he concluded they must be the remains of an American Paleolithic, one roughly comparable in age and evolutionary grade to—and therefore likely descended from—European Paleolithic groups.

Abbott's work soon caught the eye of Harvard's Frederic Ward Putnam (1839–1915), who provided financial aid, moral support, and scientific respectability at a time when Abbott badly needed all three. In 1875 Abbott was appointed as field assistant at the Peabody Museum of Archaeology and Ethnology at Harvard University. The position accorded more respect than money. Abbott was glad enough for the respect, but he certainly could have used more money: "Got a letter from Putnam. What could I not do, had I more money?" (Abbott Diary, February 7, 1876, Charles Conrad Abbott Papers, Princeton University [CCA/PU]).

Even though Abbott was a University of Pennsylvania-trained physician, by all accounts—including his own—he had a dreadful bedside manner and could not earn a living as a physician (Aiello 1967:210–11, Anonymous 1887:548).[1] But a living had to be made, for although he came from prominent, landholding New Jersey Quakers, his particular branch was not wealthy enough to sustain him. Abbott therefore took a halfhearted stab at farming and the occasional odd job—manufacturing chemicals, among others—to support himself and his family, all the while writing popular books and papers on natural history and archaeology.

It would take nearly three decades, however, before Abbott's efforts would pay off—literally—in the form of a full-time position in archaeology at the University Museum at the University of Pennsylvania. It was a position Putnam had helped secure for Abbott and thought a fitting reward for his many years of archaeological work. It was a position Abbott desperately needed. Yet, Abbott's tenure in that position would not last more than a few years. He would be released, the victim of his own irascible personality and of the explosion over the American Paleolithic, which scored a direct hit on his own work in the Delaware valley, leaving him embittered, defensive, and again unemployed.

Charles Abbott and the Archaeology of the Delaware Valley

Abbott's archaeology career began with a report in 1872 on "The Stone Age of New Jersey," in which he inauspiciously catalogued the artifacts in his collection, ascribing them to the "once mighty tribe" of the Lenni Lenape who lived across the state of New Jersey (Abbott 1872:144). There were hundreds to be described from "these people of the stone age,—these Indians, if you choose—a people who had at no time a knowledge of metals" (Abbott 1872:145–46).

Yet, among the collection were also pieces that were decidedly more crudely wrought than the remainder (he estimated the number at less than 2 percent of his sample of ten thousand) (Abbott 1876d:248). These, in fact, resembled European Stone Age implements from England, France, and the Danish "kjokkenmoddings" (Abbott 1872:153, 156–57). On this point Abbott cited the good authority and illustrated volumes of Sven Nilsson (1868), whose *Primitive Inhabitants of Scandinavia* had appeared in English just a few years earlier, and John Lubbock, Nilsson's translator and the author of the immensely popular *Prehistoric Times* (1865), to whom Abbott had sent some American specimens for his judgment (Abbott 1872:146, 153–57, 199–200, 220–21).[2] After it was published, John Evans's (1872) *Ancient Stone Implements of Great Britain* became Abbott's primary source for European Paleolithic analogues.

Abbott saw two possible explanations for these cruder forms: either "there were many execrable workmen among their tool makers; or [their] age . . . far exceeds that of the finely wrought relics." Both the crude and finely made types were "found on the surface, yet we can scarcely imagine that a people who could fashion the latter, would deign to utilize the former." Abbott concluded their comparative crudity must reflect relative age: "Take a series of whatever class of relics you may, there is always a gradation from poor (primitive) to good (elaborate), which is an indication, we believe, of a lapse of years from very ancient to more modern times, from a palaeolithic to a neolithic age" (Abbott 1872:146). Were that the case, however, Abbott realized there ought to be additional lines of evidence indicating that the two classes of artifacts were of different ages. Based on stratigraphy, context, and association, it appeared they were.

Although the "rude implements" were occasionally found on the surface with Indian implements, Abbott insisted the two were not the same age since the rude forms routinely occurred "deeper in the soil than the majority of the so-called 'surface' specimens" (Abbott 1876d:247). Indeed, where "one paleolithic implement is found upon the surface, a hundred are quite deeply imbedded [*sic*] in the soil, and in the underlying gravels" (Abbott 1876a:52).

The question of why *any* of the ostensibly older crude implements were found on the surface at all seemed easy enough to answer: the older artifacts had been exposed by geological processes, or picked up by later Indians who viewed the pieces as "venerable relics of a departed people" and used them for their own purposes. In either case, the difference between the rude and Indian artifacts was easy enough to spot, since the rude relics "invariably exhibit a greater degree of weathering" (Abbott 1876a:51, 1876b:66).

Similarly, Abbott observed that the rude implements "are never found in

[Indian] graves, or in any graves that we have examined," which would be expected were they still being made and used in recent times by Indians (Abbott 1876a:52). Finally, he pointed out that among the rude implements there were certain forms, notably, what he called "turtle-backs," which were not usually associated with the tools of Indians, just as there were "common Indian relics" that never occurred deep in the soil alongside the turtlebacks. That seemed good evidence that the rude implements were not just Indian implements that had tumbled to these depths from the surface, for if they were, one would also expect other "Indian [stone] relics" as well as pottery to occur in the gravels (Abbott 1876c:332; 1877a:32, 37).

Abbott was not alone in seeing a similarity between these relics and the crude implements of Paleolithic Europe. When he made the first of his donations to the Peabody Museum in 1872 (some eight hundred specimens), curator Jeffries Wyman readily spotted "several implements which, as Dr. Abbott states, very closely resemble the celts of the drift period of Europe, especially those found at St. Acheul, two or three of which, except for their material, could hardly be distinguished from them" (Wyman 1872:27).[3]

Early on, Abbott envisioned a straightforward historical (he called it "evolutionary") sequence from those who fashioned the crude implements to those who made the more finely polished ones. In the rude turtlebacks, hatchets, and spears, he saw the early steps of "an unbroken line of development in the manufacture of tools" leading up to the more modern stone weapons (Abbott 1876d:252). The rude implement makers, he believed, were ancestors of the Indians (Abbott 1872:147, 153, 1881a:124). "Whatever the origin of the American aborigine . . . there can be no question as to his *condition* from the date of his first appearance on American soil to the time of the arrival of the European settler, a period of immense duration, during which the puzzling red man passed from the Paleolithic to a Neolithic condition" (Abbott 1876d:380, emphasis in original). All this was in keeping with the "theory of the gradual progress of mankind," amply demonstrated in Europe, "although the older Stone age of the American race or races does not date back as far into prehistoric ages as is probable in other continents" (Abbott 1876d:380).

Still, it might get close to being that old. In 1873 Abbott reported his first discovery of three deeply buried and crudely fashioned artifacts from the "river drift" of the Delaware valley. These "drift implements" were distinct from the rude artifacts found on the surface and associated with ordinary "Indian relics," and even more distinct from the Indian relics themselves. The distinction, however, was based more on stratigraphy than form: the drift implements were "wholly different . . . inasmuch as all three were taken from this gravel at great depth, and all *beneath undisturbed layers of fine sand.*" Because of their position below several undisturbed strata and at great depth, these "true drift implements [must have been], fashioned and used by a people far antedating the people who subsequently occupied this same territory. . . . and we must admit the antiquity of American man to be greater than the advent of the so-called 'Indian.'" Abbott's observations on the stratigraphic position of the drift implements, however, were not based on excavations of material in place but rather on the collection of artifacts as they eroded out of sand and gravel deposits exposed in the bluffs and banks of the Delaware River and its tributaries (Abbott 1873:205–6, 209).

As his collection of drift implements grew through the mid-1870s, Abbott became increasingly uneasy with the idea of a historical or evolutionary continuity between these and the more elaborate artifacts of the Indians. Yet, his uneasiness seems to have been motivated less by any theoretical concerns about deriving one population from the other than from the simpler observation that the Paleolithic implements were significantly "different from the others in many respects" (Abbott 1876b:72). He could not imagine how one could become the other. The Paleolithic artifacts were different in their ruder form and finish, were made of a distinctive raw material type (later identified as argillite), had a greater degree of surface weathering, and included artifact types—such as the turtleback—that were not seen in later, Indian assemblages (Abbott 1876c:330–32). Ultimately, in early 1876, he completely abandoned the idea of "an unchecked development, a gradual merging of the one into the other condition," believing instead these were "traces of distinct peoples" (Abbott 1876b:64, 66).

The two, of course, differed in age as well. While Abbott no longer saw them as historically or evolutionarily related (in terms of the two being related as ancestors-descendants), the difference in their artifacts still followed the *general* evolutionary principle that there was a "progressive" and "gradual improvement" of artifact forms in time. Some are poorly made, others not, and oftentimes those that are poorly made are found deeper in the soil. As Abbott summed it up: the "nearer the surface, [the] finer the finish" (Abbott 1876b:71). Indian remains might be old, but on morphology and depth the drift implements had to be older still.

Abbott estimated the precise age, in years, of the Indian remains by calculating (or at least assuming) rates of artifact change and soil buildup atop sites in the forest and in river valleys. So, for example, he argued that the crudest form of Indian arrowhead in his collection was about 1,300 years old, based on the accumulation of forest litter. But since that particular class of arrowhead was already at the third of four "degrees of excellence" in the evolution of arrowhead making, a first-degree (or the earliest) Indian arrowhead had to be at least 2,600 years older, hence slightly over 3,000 years old. All things considered, then, "from thirty-five to forty centuries ago [3,500–4,000 years], at least . . . the Indian appeared in what is now New Jersey" (Abbott 1876b:72).

The drift implements were obviously older than that. How much older in absolute years was more difficult to estimate. Certainly, their presence in the drift showed "that not long after the close of the last glacial epoch man appeared in the valley of the Delaware. Given that Abbott then followed James Croll (1875) and James Geike (1874), who, on the basis of astronomically driven changes in incoming solar radiation, put the end of the glacial period at eighty thousand years ago, he assumed a substantial antiquity for the initial occupation of the Delaware valley (Abbott 1876c:330, 335).[4]

If these drift groups were unrelated to Indians, then what became of them? Since Abbott found drift implements near the surface, he believed their makers occupied the Delaware valley up until the initial wave of Indians arrived (who, according to Abbott, were still fashioning relatively—for them—rude artifacts and had not yet "advanced" to the "Neolithic condition"). Given the evidence Abbott saw in Indian tradition and history of their "being a usurping people," he thought it likely that upon their arrival they drove off the descendants of the makers of the drift implements (Abbott 1876b:72, 1876c:330, 333).

But where were they driven? Abbott did not think it probable that they should have entirely left the country, so he looked among neighboring groups for their descendants. He found them in the Eskimo, although he didn't go there directly from the Delaware valley. Instead, he took his argument on a detour through Paleolithic France.

Abbott was struck by the similarity he saw between "the Delaware Valley implements [and] those of Europe . . . [especially] the relics from French Caves" (Abbott 1876c:334). The Paleoliths of New Jersey, in his view, "exactly reproduce" artifact types of the Reindeer People of the French Caves, whose artifacts were illustrated in Edouard Lartet and Henry Christy's just published (1875) *Reliquiae Aquitanicae.* Lartet and Christy had observed the apparent similarity between European Paleolithic artifacts and those still in use among the Eskimo (for example, Lartet and Christy 1875:16, 25). Abbott merely closed the syllogistic loop. "If, therefore, the rude implements of the Delaware Valley gravels resemble those of the caves of France, and the French troglodytes were identical (?) with the Eskimo, it is fair to presume that the first human beings that dwelt along the shores of the Delaware were really the same people as the present inhabitants of Arctic America" (Abbott 1876c:334).

On the whole that argument met little resistance, although the United States National Museum's Otis Mason gently complained of Abbott's "too hasty adoption of the generalizations of some English and French archaeologists, with reference to the order of culture on our continent" (Mason 1877:310; see also Abbott 1877b).

All things considered, though, it was not a bad start. As the archaeologists of the European Paleolithic had done before him, Abbott had seemingly shown that there was a class of artifacts in America that were by their morphology, context, association, and stratigraphic position distinctive from those of the recent Indians. He had shown that those artifacts came from drift deposits, possibly dating back to the retreat of glacial ice, and bore a striking resemblance to artifacts of similar age in Europe. Finally, through his Eskimo-connection, he had provided a framework for the origin and subsequent history of these American Paleolithic groups, which linked them directly to the prehistory of Europe.

Abbott and Putnam

The first of Abbott's papers (which was non-archaeological) was published in 1870 in a new quarterly, the *American Naturalist.* The journal had been founded four years earlier by a breakaway group of Louis Agassiz's Harvard University students who were dismayed by Agassiz's staunch anti-Darwinian stance and wanted an outlet for works on evolution (Mark 1980:19). Frederic Ward Putnam, then at the Peabody Academy of Science in Salem, Massachusetts, was one of those mavericks. Putnam and Abbott's acquaintance on the ground of natural history quickly converged on their common interest in archaeology, and Putnam soon began receiving boxes of artifacts Abbott was collecting from his ancestral farm, Three Beeches, in the Delaware valley just outside of Trenton, New Jersey.[5] Putnam nurtured and encouraged Abbott's archaeological work, and in turn Abbott sent his early papers on archaeology to Putnam's *American Naturalist.*

Naturally, the attention from Putnam was welcome, and their association began in earnest in 1876 when Putnam and his family, on their way to the Centennial Exposition in Philadelphia, stopped off in Trenton to see Abbott. Although he was by then quite familiar with the artifacts from the region, this was Putnam's initial opportunity to see firsthand the fields where Abbott had collected them. For his part, Abbott welcomed the visit, admitting he "was too excited for my own good" (Abbott Diary, September 1, 1876, CCA/PU).

Two days later, Abbott and Putnam were climbing down the bluffs along the Delaware, just below Trenton's Riverview Cemetery, looking in the gravel for pre-Indian traces. Three artifacts were found. Abbott recorded that he "was as positive to [their] paleolithicity as Putnam was cautious" (Abbott Diary, September 3, 1876, CCA/PU).

The following day, Abbott went with Putnam to Philadelphia, where they toured the Exposition and participated in a Centennial Convention of Archaeology. By week's end, Putnam was exhausted and Abbott was thoroughly bored. Abbott's interest in archaeology, he frankly admitted, did not extend beyond the edge of the Delaware valley, and he wearied of this "one subject [archaeology] that has been uppermost for so many days." So much talk about archaeology was Putnam's business as curator of Harvard's Peabody Museum, and not his own, Abbott grumbled (with more than a touch of envy), since for him archaeology was "not in any sense professional" (Abbott Diary, September 6–11, 1876, CCA/PU). On that point, the week in Philadelphia weighed heavily on Abbott's mind.

Philadelphia was an acute reminder, when Abbott needed none, of both his borderline status in the field and his own financial precariousness. While others spoke in "glittering generalities" about archaeology on a wide canvas, Abbott could reply only of the Delaware valley, though he quietly sought comfort in the knowledge that "I am only the only man living [who] knows anything about it" (Abbott Diary, September 7, 1876, CCA/PU). While those who could speak to a broad range of issues were among the ranks of professional scientific workers, Abbott saw clearly that his "living must come from the farm and can never come from any scientific activity" (Abbott Diary, September 8, 1876, CCA/PU). But even the farm work was unreliable: "The lack of capital makes me unequal to carrying out the business of a farm, especially as the expenses are considerable, due to losses at the outset, and the fact that I am not physically able to do what is known as farm work. . . . As to my scientific tastes, they must be curbed, as they do not bring in any revenue. It is a hard lot, but I cannot change it. It is grin and bear it to the end" (Abbott Diary, September 13, 1876, CCA/PU).

Abbott had inherited his ancestral family home, Three Beeches, in 1874, and that provided some financial relief. But throughout the late 1870s and 1880s he was "bothered all the time by money matters" and suffered periods in which he felt like giving up archaeology altogether (Abbott Diary, June 14 and 18, 1878, CCA/PU). On occasion, he did, and sent Putnam his woeful cries: "Forced out of the ranks of scientific workers, of course you will all very soon forget me, but I have one request to make. Please do not erase my name from the list of recipients of your Annual Reports. It will be a pleasure for me to yearly note your progress" (Abbott to Putnam, November 20, 1881, Peabody Museum Papers, Harvard University [PMP/HU]). But even when Abbott's enthusiasm for the subject returned, and it always did, he could not escape the fact that archaeology was "wholly unproductive of pecuniary reward" (Abbott to Wright, March 3, 1882, George Frederick Wright Papers, Oberlin College Archives [GFW/OCA]).

Putnam provided Abbott a measure of scientific respectability, and even a bit of financial help. But only a bit: between June 1876 and January 1879, Abbott received from the Peabody Museum a grand total of $583.25 for his explorations in the field, collections, and a brief stint of winter museum work—far less than the peripatetic Edward Palmer received over that same period. Putnam may have been impressed by what Abbott had done and was doing, but when it came to channeling the Peabody's scarce dollars, relatively few of them went to Abbott.[6]

That wasn't because Abbott's work was outside the scope of the Peabody's charter. In fact, when the museum was established a decade earlier, George Peabody expressly refrained from putting conditions on his bequest, save that "in the event of the discovery in America of human remains or implements of an earlier geological period than the present, especial attention should be given to their study, and their comparison with those found in other countries" (Peabody 1868:26). That suited Putnam, who was interested in the question of human antiquity in America.

Putnam gave Abbott a few token payments in return for work already done, to support his report writing for the Peabody Museum, and perhaps to encourage Abbott's future loyalties, should he subsequently make any further significant discoveries. It worked; Abbott occasionally had to apologize for being unable to send specimens because he was on a "retaining fee" from the Peabody Museum (e.g., Abbott to Wright, March 13, 1882, GFW/OCA). Several years later, however, Abbott would take a darker view of that relationship. The Peabody Museum had already "squeezed the juice from the lemon, and now throws the skin away. I suppose I am only the skin, now, to them" (Abbott Diary, July 13, 1883, CCA/PU).

Even so, Abbott and Putnam needed each other: Abbott to provide Putnam with essential archaeological collections and evidence, Putnam to cloak Abbott's collecting efforts in scientific respectability. But it was not to be a harmonious union. In fact, Hinsley (1985) explains, the marriage was doomed from the start. As the manager of Abbott's Trenton Paleoliths—if these were indeed the traces of the earliest Americans—Putnam and the Museum could bask in the reflected glory. But Putnam also had much to lose if the Trenton evidence went sour. At Harvard he was responsible for the scientific reputation of a discipline and a museum, and had his eye on the endowed yet unfilled Peabody Professorship of Anthropology. He was naturally anxious to tread cautiously, for he'd suffer on several counts if work he endorsed proved to be scientifically unsound (Hinsley 1985:61; Mark 1980:31).

Abbott was as ambitious as Putnam was cautious, and as a result their association was rocky from the start and would grow ever more so if Abbott was left to his own, for he demonstrated he was quite willing to express his opinions about the *meaning* of the artifacts from the Delaware gravels (Palmer never wrote about what *his* collections meant, which may have had some influence on the greater degree to which Putnam supported his activities). Facing the prospect that a vocal, unchecked, and ambitious Abbott might embarrass him, Putnam sought—using the leverage provided by his financial support and the status Abbott gained by his affiliation with the Peabody—to curb Abbott's enthusiasm and guide his work to more substantial scientific ground. That was not Abbott's style, and he sputtered almost constantly: "I sent to Putnam what I believe to be four Paleolithic implements. I think Putnam is skeptical still in the subject, but I am not" (Abbott Diary, December 14, 1876, CCA/PU).

While Putnam may have been reserved about Abbott's work, others were less so, particularly the Reverend George Frederick Wright (1838–1921), an Andover, Massachusetts, theologian and an up-and-coming figure in glacial geology. Wright and a young Philadelphia geologist, Henry C. Lewis (1853–88), visited Trenton and helped put the geology of the gravels on firm footing.

There were matters left unsettled: the archaeological discoveries were on the cusp of an emerging discussion in geology over the number and timing of glacial advances in the Americas. But whatever the outcome, it seemed evident to Wright that Abbott's American Paleolithic was glacial—comparable in age and grade to the European Paleolithic. And Abbott had that on no less authority than the eminent British prehistorian Boyd Dawkins, who visited Trenton in late 1880 and spent three "glorious" days with Abbott "relic hunting" in the gravels. Many Paleolithic artifacts were found, to which Abbott crowed, "*Dawkins gives as his opinion that I have made, unquestionably, the discovery of paleolithic man in the Delaware Valley. He enthusiastically endorses my position, and I need no greater authority to express my opinion*" (Abbott Diary, November 19, 1880, CCA/PU, emphasis in the original). As Dawkins himself later wrote: "The implements [at Trenton] are of the same type, and occur under exactly the same conditions [and were the same age], as the river-drift implements of Europe," clearly establishing that the "river-drift men" had spread around the globe in Pleistocene times (Dawkins 1883:343, 347–48).[7] He wasn't alone in his praise. In early 1881 Abbott's Paleolithic archaeology of the Delaware valley took center stage at the Boston Society of Natural History, and the city's scientific elite were there to bear witness to his findings and testify to the reality and antiquity of the American Paleolithic.

Status and Rank

In ten years Abbott had gone from Trenton to Boston, and in so doing traversed the even longer distance from the periphery of archaeology to near its center. In recognition of his accomplishment he would be lionized at home and abroad as America's Boucher de Perthes. Abbott's book *Primitive Industry* (Abbott 1881b) was likened to Evans's *Ancient Stone Implements of Great Britain* (1872), and the Trenton gravels were given the same pride of place in American archaeology as Abbeville in European archaeology (for example, Anonymous 1882:37, Putnam 1888:421, Topinard 1893, Wallace 1887, Wilson 1893). But Abbott's professional triumph did little to overcome his financial insecurity. In those years there were ample opportunities to contribute to archaeology but few chances to benefit from it, and Abbott's personal situation remained precarious (Hinsley 1985:68). It was as though he never left Trenton.

Still, Abbott had boldly taken the elusive idea of a deep human antiquity—abandoned by others as unresolvable—and revitalized it. It was Abbott who, with Putnam and Wright, saw that the future of American archaeology appeared to be deep in the past—perhaps as much as ten thousand years. This was far older than any other known human traces in North America, and certainly far older in relative and absolute terms than Abbott himself had imagined just ten years earlier when he first noticed the artifacts that appeared different from those of the Indians.

Abbott was getting greedy too: soon even ten thousand years would not be enough. For with only ten thousand years of prehistory, Abbott complained, he would be "compelled to crowd several momentous facts in American archaeology into a comparatively brief space of time." With only ten thousand years, he would have to conclude that "paleolithic man was essentially a coast ranger" who would not have had the time to penetrate far inland before being "met by southern tribes, who drove them northward, exterminated or absorbed them." Alternatively, and he clearly favored this argument, if the Paleolithic groups were related to the Eskimo, then "man was pre-glacial in America" and was driven southward (and subsequently retreated northward) in consort with the ice advance. There was not then any direct evidence for such preglacial humans, but Abbott was confident that it would ultimately be found (Abbott 1883:359).

Abbott's confidence was as high as his archaeological stock—which was rising even outside the Trenton-Boston circuit in which he traveled. In *Science,* John Wesley Powell (1883), director of the Bureau of Ethnology in Washington, D.C., formed several years earlier as an anthropological research wing of the Smithsonian Institution, took a favorable notice of Abbott's work, accepting his claim to have found evidences of the existence of an American Paleolithic.

Yet, even while Abbott's archaeological stock rose, his personal and occupational fortunes hardly improved. He came to realize that however much his archaeological discoveries were valued by the community—and they were—that value would not easily translate into financial security or professional standing. Abbott desperately wanted both. He coveted the academic life he experienced during his occasional visits to Cambridge to work on his collections. After one especially memorable evening spent roaming the intellectual landscape with several of Harvard's faculty, Abbott desperately wished he lived in such an environment and not Trenton—a "purely commercial and fanatical neighborhood" (Abbott Diary, November 25, 1876, CCA/PU). But the best he got from Putnam was an appointment as an unsalaried, part-time "Assistant in the Field" of the Peabody Museum. There was little more

Putnam—or anyone else at the time—could offer. Forced to find employment, Abbott ultimately found himself trapped as a badly misfit clerk in a Trenton bank, and increasingly isolated from archaeology. It was not the position or the status to which he aspired: "I wish to heaven I could be satisfied to be a clerk and a nobody, but I cannot" (Abbott Diary, December 4, 1883, CCA/PU). His days at the bank were tedious and lifeless, and even bankers' hours left little time for writing or searching for Paleoliths, though plenty of time for feeling sorry for himself: "Up town and at bank until 3 PM, which seems to wholly use up the day; as I cannot get up early enough to write before breakfast, and am too lazy after I get home" (Abbott Diary, July 13, 1883, CCA/PU; this diary entry is typical; see also Hinsley 1985). Putnam was no help at all, Abbott grumbled, since he could do "nothing but try to cheer with hopes of better times and such stuff," and Abbott put no faith in such hopes (Abbott Diary, July 13 and 16, 1883, CCA/PU).

In his unhappiness, Abbott convinced himself Putnam and the Peabody Museum had strung him along all those years and were ready to throw him overboard now that they had his collection. Abbott's real problem with Putnam was, of course, less insidious: he simply needed constant attention and assurance. This Putnam gave in some measure, but it was never enough for Abbott: "He [Putnam] cannot find time to answer my brief questions on matters that used to be of importance to him" (Abbott Diary, November 27, 1883, CCA/PU. See also Hinsley 1985:65). To make matters worse, Putnam insisted Abbott not stray in search of archaeological affection. The Smithsonian Institution's secretary, Spencer Baird, had written Abbott asking for samples of the Trenton Paleoliths, but when Putnam got wind of the request he immediately urged Abbott to do "nothing archaeologically in connection with any other museum than the one at Cambridge," appealing to his loyalty and holding out the promise of future reward. Little though it was, that was enough to soothe Abbott's deep insecurity: "He [Putnam] has much to say about 'hope,' and 'looking forward' &c and that the time may come when the fact that I have stuck to the museum, will result in the museum 'sticking to me.' . . . I do not expect to be able to accomplish any work of importance, while tied down to my present position, but I will take Putnam's advice and be an enthusiastic 'hoper'" (Abbott Diary, December 8, 1883, CCA/PU).

Abbott would pay a small price for his loyalty to the Peabody, which was then in the midst of a minor skirmish with the Smithsonian over the rich archaeological fields of southern Ohio.[8] When he later wrote Baird for a copy of the Smithsonian's latest *Contributions to Knowledge* volume, Rau's

"Prehistoric Fishing in Europe and North America," he was refused outright, ostensibly (according to Abbott) because he had not sent Baird any Paleolithic implements. "I'll see him dammed first," an angry Abbott wrote in his diary: for the moment, he would stand by Putnam (Abbott Diary, April 22, 1885, CCA/PU).

And Putnam stood by Abbott. Mostly. In the summer of 1884, rumors were flying in the Philadelphia scientific community that something was amiss at Trenton. University of Pennsylvania paleontologist E. D. Cope wrote Putnam to report that a local grammar school teacher was claiming he had accompanied Abbott in the field and seen him "*bury* the paleolithic 'finds.'" Putnam did not know what to make of the charge; Cope urged him to look into the matter: "My prepossessions have always been in favor of Abbott, but there is no doubt he has a lively imagination, & the charges appear to be—some of them—well founded. You should look them up privately first & when you, as grand jury, find an indictment, then bring him to face his accusers" (Cope to Putnam, May 20, July 7, and July 11, 1884, Frederic Ward Putnam Papers, Harvard University [FWP/HU]). Putnam made discreet inquiries and apparently satisfied himself that Abbott was innocent of the charges. Still, it was another warning signal that he had better keep a close watch on Abbott and be prepared to distance himself quickly if Abbott fell into ill repute or started to tarnish Putnam's and the Peabody's scientific image.

Abbott sensed the change in Putnam's attitude and resented it. In September of 1884, Putnam and Abbott had a more unpleasant than usual exchange over Abbott's relationship to the Peabody Museum. Abbott's letter so offended Putnam that he burned it, then admonished Abbott to write again "when in [his] right mind" (Abbott Diary, September 30, 1884, CCA/PU). Abbott was furious: Putnam, he fumed in his diary, cannot "reconstruct me into an enthusiastic local archaeologist."

So Abbott began to court Daniel Garrison Brinton (1837–99) at the University of Pennsylvania. Abbott's efforts to "re-unite with Philadelphians" continued over the next year or so, yet bore little fruit. Indeed, in 1886 those efforts slackened as his relations with Putnam again began to improve. By then he had quit the bank, his latest nature books (*A Naturalist's Rambles about Home* [1884] and *Upland and Meadow* [1886b]) were published to critical and even financial success, and Abbott could afford to feel a bit more generous. Even better, Putnam's Peabody Museum *Annual Reports* for 1884 and 1885 had appeared in early 1886, and in them he bestowed on Abbott alone "the credit of having worked out the problem of the antiquity of man on the Atlantic coast" and heaped praise on the Peabody's Abbott Collection as one of the most instructive in the museum. Numbering as it now did over 20,000 pieces (Anonymous 1886, 1887), it was certainly one of the largest in the museum. With it, Putnam could display the specimens from the three prehistoric periods (the Paleoliths from the Trenton gravel, the intermediate artifacts, and those of the New Jersey Indians), "superposed in time, arranged in sequence side by side, and in such numbers that the cumulative evidence, always asked for by the archaeologist, is amply presented."

Just months earlier *Science* (Anonymous 1886) had devoted several columns to the Abbott Collection, calling it "one of the most important series of the kind ever brought together, and one which archaeologists will consult for all time to come," and heaped praise on Abbott's "industry and sharp-sightedness." Given the echoes in *Science* of Putnam's own description and praise of the Abbott collection, Putnam may have had an indirect hand in this anonymous report, or perhaps wrote it himself (although that seems less likely). However direct or indirect his influence, Putnam exploited the promotional windfall, reprinting it in its entirety in the Peabody Museum's *Annual Report.* Here, for the museum's trustees and donors among the Boston Brahmins, as well as for the president and board of Harvard College (which into the mid-1880s did not even mention the museum in its own *Annual Report*), was independent testimony of the valuable research being carried out under his direction. The oldest human remains in America may not equal the high art of classical antiquity that was especially attractive to Boston's philanthropists, but they could at least rival the best of European prehistory (Putnam 1886a:408, 1886b:487, 491; also, Hinsley 1985).

For Abbott the *Science* report, but especially Putnam's words (whatever their deeper motive), were extremely soothing, and he returned the compliment, writing a flattering biographical sketch of Putnam for the September 1886 *Popular Science Monthly* (Abbott 1886a). Shortly before, Putnam's own stock had risen considerably, when he was nominated to the Harvard position he had so long coveted: on January 12, 1887, he was appointed Peabody Professor of American Archaeology and Ethnology (a position that was endowed in Peabody's original 1866 gift but had remained unfilled). It had not been an easy prize. Serious consideration of Putnam's candidacy did not begin until 1885, and then for nearly two years his path was blocked by Harvard's Alexander Agassiz. The son of Louis Agassiz, and himself a geologist who inherited the directorship of Harvard's Museum of Comparative Zoology, Alexander shared his father's low opinion of Putnam (it seems unlikely there was any lingering family resentment over Putnam's defiance of the senior Agassiz twenty years before, although it is perhaps worth noting there was the potential for it: in 1865 Louis Agassiz spurned a sizable gift from George Peabody, who then turned the money over to Agassiz's rebellious student at Salem [Putnam] to endow the Peabody Academy of Science in Salem) (Hinsley 1985:49–50, 61; Peabody 1868:27).

Just as Abbott's work in the Delaware valley was gaining prominence, so too was the American Paleolithic, which began in the mid-1880s to assert itself in other parts of the country. Scattered reports were coming in of other Paleolithic sites from the eastern seaboard into the upper Midwest. Most of these sites had artifacts like those of Trenton. Some were found under geological circumstances—such as an association with glacial features—that suggested an antiquity comparable to Trenton. A few hinted at a still older Paleolithic human presence, perhaps even dating to an interglacial epoch (which itself was controversial—not just because of the archaeological implications but because it became embroiled in debate within the geological community over whether there had been multiple glacial periods).

The American Paleolithic was rapidly spreading, and spawning a formidable literature testifying to the foresight of Abbott's vision. By the late 1880s, the American Association for the Advancement of Science (AAAS), the Boston Society of Natural History, and the Anthropological Society of Washington had hosted a series of papers and symposia testifying to a deep human antiquity (for example, Mason et al. 1889, McGee 1888, Putnam et al. 1888, 1889, Wright 1889, see discussions in Meltzer 1983, 1994). The question was no longer whether humans had reached the Americas during the latter part of the Pleistocene, as at Trenton, but just how much farther back in time they may have arrived. It was a remarkably smooth coming of age—better than could have been hoped for a decade earlier by even the most optimistic champions of a deep human antiquity.

Abbott rode the crest of that wave. In the late 1880s, he was accorded the singular honor of election to the vice-presidency of section H (anthropology) of the American Association for the Advancement of Science. In his vice-presidential address to the section H, he could hardly keep from gloating. But then he felt he'd earned it. For the moment, then, Abbott was doing very well indeed. At least on the archaeological front.

Abbott's Triumph—Abbott's Troubles

The year 1889 would prove a banner one for Charles Abbott, though it started out badly. He had been vice-president of section H the year before, but it now appeared as though problems on the family farm, a failure to receive expected funds, and other financial pressures were conspiring to prevent Abbott from even attending the 1889 AAAS meetings in Toronto. He desperately wanted to go, for his research on the "argillite matter" (which he supposed was a "pre-Indian" and post-Paleolithic "culture") was entering a new phase, and though he had worked on it longer than anyone else, he feared others might soon steal his thunder. Wrenching his emotions still further were "tempting *bribes*" from Washington and Philadelphia that severely tested his loyalty to Putnam and the Peabody Museum. In the "bitterness of painful disappointment," Abbott wrote Putnam, like a "drowning man catching at a straw," and poured out his heart. Putnam's reply soothed Abbott's feelings, and apparently promised to personally help pay Abbott's expenses to Toronto (Abbott to Putnam, June 21, 1889, FWP/HU). In obsequious tones uncharacteristic of Abbott, he gratefully assured Putnam of his continued loyalty:

> I have never for a moment thought of doing otherwise than I have done. *You* made it possible for me to become an archaeologist; and as I recently wrote to [Henry] Henshaw "every jot and tittle of *ideas* as well as specimens, rightfully belongs and certainly shall go to the Peabody Museum."

But he still felt beaten:

> the older I grow, the more I believe myself a fool to try, being poor, to take a prominent part in the scientific world. Money makes science, as well as the man, go [Abbott to Putnam, June 21, 1889, FWP/HU].

Putnam, having successfully secured his own position, would work to secure one for Abbott as well, and in a few short months in late 1889, all the years of frustration, penny-pinching, and financial and intellectual insecurity for Abbott were forgotten. Over October and November of 1889, Abbott was deep in negotiations with the University of Pennsylvania Museum for a salaried position in archaeology. Putnam was quietly helping behind the scenes, explaining to William Pepper, provost of the university, just what the Peabody Museum's arrangements were with field assistants like Abbott, how much salary Abbott received (none, only expenses, which Putnam duly exaggerated in favor of Abbott), and what kind of person Abbott was: "a singular man, and one who has to be treated differently from common mortals and as I [understand?] him and thoroughly appreciate him I have not put him under any obligations that he did not prefer himself."

Putnam urged Pepper to appoint Abbott to a full-time position, not a part-time one. Otherwise, Abbott would "get into an unhappy state of mind by hopes deferred" and (Putnam not-so-subtly threatened) likely keep his present arrangement with the Peabody Museum: "I have told him that I think I should not accept any secondary position, for if that is all he is to have we can do as well for him here." Putnam assured Pepper that Abbott would work hard for his new museum and be an important man for its development (Putnam to Pepper, November 8, 1889, University of Pennsylvania, Museum Archives [UPM]).

On December 1, 1889, Abbott became the University Museum's first curator of archaeology, at the munificent salary of one thousand dollars a year—the only salaried curator (Madeira 1964:20; Abbott Diary, October 23 to December 2, 1889, CCA/PU). It was, Putnam thought, a fitting reward. "Dr. Abbott was for years placed in a very unpleasant position by the non-belief of many persons in the great discovery which he made showing that man existed in the Delaware Valley at a time preceding the deposition of the Trenton gravel. . . . Now that the scientific world gives him the full credit he so richly deserves, and he is offered an honorable position by the University of Pennsylvania, I am filled with happiness for his sake" (Putnam to Pepper, October 28, 1889, quoted in Hinsley 1985:65).

At the annual meeting of the Peabody's board of trustees, Putnam asked the board to pass resolutions of thanks to Abbott for his fourteen years of service and collecting as a special assistant in the field, to specify that the nearly thirty thousand specimens sent in by Abbott thereafter be known as the Abbott Collection, and to convey to Abbott and the University of Pennsylvania the Peabody board's congratulations on the hire (Putnam 1889:71–72). The resolutions passed.

Putnam thought the worst of Abbott's troubles were behind him. Actually, they were just beginning.

In November of that year, William Henry Holmes (1846–1933) of the Bureau of Ethnology announced—based on work at the Piney Branch quarry in Washington, D.C.—that artifact *form* had no chronological significance whatsoever. He arrived at that conclusion in a relatively straightforward fashion. As Holmes saw it from the quarry debris littering Piney Branch, the process of artifact manufacture comprised three stages, which transformed a simple cobble through "successive degrees of elaboration" into long leaf-shaped blades. During the first stage, the cobble was flaked by percussion on one side, thereby producing "a typical turtle-back." It was then turned over, and the other side was flaked, resulting in a rough bifacial preform. In the last stage the biface was thinned and made symmetrical, its edges were straightened, and it was roughly pointed and occasionally a stem or haft was put on. There, with the "neat, but withal rude, blades," the process at the quarry itself ended. The blades had "graduated from the school of direct or free-hand percussion" and were then carried off the site where "final finishing" (by indirect percussion) took place and the artifacts were used. Naturally, this explained why the rude forms rarely (if ever) showed signs of use: they were neither finished, nor even intended to be used, and if they had been used it was only because they had been pressed into "emergency" service (Holmes 1890:12–13, 15–16, 17).

For Holmes, the process he reconstructed at Piney Branch, and which he had been able to replicate experimentally, ran along a very narrow and well-defined pathway: "every implement resembling the final form here described made from a bowlder [*sic*] or similar bit of rock must pass through the same or much the same stages of development, whether shaped to-day, yesterday, or a million years ago; whether in the hands of the civilized, the barbarian, or the savage man" (Holmes 1890:13–14).

That singular process "left many failures by the way." Would-be artifacts were discarded because they broke during the flaking process, had unmanageable features (were too thick to flake or had humps that could not be removed), or were otherwise flawed. *Everything* at Piney Branch was therefore "mere waste" or a "failure" of the manufacturing process. That included the allegedly Paleolithic turtlebacks. "It causes me almost a pang of regret at having been forced to the conclusion that the familiar turtle-back or one-faced stone . . . together with all similar rude shapes, must, so far as this site is concerned, be dropped wholly and forever from the category of [finished] implements" (Holmes 1890:14).

He said *almost* a pang of regret, but he was just being polite. He didn't seem to mean anymore by it than his apparently modest suggestion that his conclusions had relevance only to Piney Branch. After all, in the very next paragraph he declared: "what is true of the rude forms of this particular locality may be true also of all similar forms found throughout the Potomac Valley" and beyond. Piney Branch may have been but a single site, but Holmes was certain it "was representative of a class, and will serve in a measure as a key to all" (Holmes 1890:3). Thus, merely because something looked primitive did not make it ancient. With that realization came the inevitable corollary: the American Paleolithic had no typological anchor and without geological evidence was essentially adrift in time.

Holmes admitted that even though Abbott's Delaware valley Paleoliths seemingly corresponded to the "failure shapes" at Piney Branch, their geological occurrence "seem[s] to make them safe indices of the steps of progress" (Holmes 1890:15). Still, even a careless reader could see that the Piney Branch work struck deep into the core of the American Paleolithic and that it would be only a matter of time before Holmes trained his sights on other Paleolithic claims.

Word soon reached Abbott of Holmes's paper, and he dashed off a letter to Henry Henshaw, editor of the *American Anthropologist:* "Will you please get for me a copy of Holmes (?) paper on the Paleolithic!!! finds near Washington?" (Abbott to Henshaw, February 20, Bureau of American Ethnology, National Anthropological Archives). After sending Abbott Holmes's paper, Henshaw followed with an invitation. Would Abbott consider coming to Washington, preferably immediately?

> My particular reason for saying immediately is that I have just returned from a visit to Piney Branch where Holmes is continuing his investigations into the old quarry sites. . . . It seems to me that you would become particularly interested in the matter since you have done so much work at Trenton, and a visit here just now could not fail to prove instructive. Mr. Holmes would be very glad to see you, and we will all do what we can to make your visit pleasant and instructive [Henshaw to Abbott May 21, 1890, CCA/PU].

"Instructive," Henshaw had said. Twice. But Abbott wasn't looking for instruction. He was looking for *confirmation.* Abbott was puzzled, perhaps a bit wary, but went to Washington to tour Piney Branch with Holmes. Not surprisingly, he was unmoved by what he saw. Holmes may be correct on the archaeological significance of Piney Branch, Abbott wrote in his first *Annual Report* to the University Museum, but only about what he'd found at Piney

Branch. Abbott would not countenance any generalizations beyond that, particularly in regard to the Paleolithic, for in Abbott's mind "the inferences drawn [by Holmes] are too sweeping, and have not necessarily the bearing upon the question of man's antiquity in America which he [Holmes] practically claims" (Abbott 1890:9).

Abbott had his reasons for drawing the line: for one, the Piney Branch failures were not at all identical to the "true paleolithic implements" of the Delaware River valley. The humps that at Piney Branch marked discarded turtlebacks were often found on Delaware Valley Paleoliths as "elaborately worked as any arrow-point" and even on more refined Indian artifacts, obviously contradicting Holmes's assertion that humped-backed implements were necessarily signs of manufacturing failure (Abbott 1890:8–9). Further, Paleoliths did not form a graded series with, for example, argillite arrow points, drills, and scrapers, testimony that the former were not merely the early stages of the latter. There was a wide gap between the two, far wider than Paleolithic critics could close.

Finally, it seemed to Abbott that Holmes was laying too much stress on the implements themselves and not enough on the circumstances under which they were found. After all, were Holmes right (that all rude implements "howsoever, and wheresoever, found, are Indian 'failures'"), it would mean removing the Paleolithic implements of Europe, Asia, and Africa from the prehistory of those continents. That was hardly warranted, for the conditions under which Paleolithic specimens were found—and that certainly included those of the Delaware valley—were wholly different from what Holmes had at Piney Branch. At Trenton, the Paleoliths were characteristic of a horizon, and by "no verbal jugglery" could they be relegated to an incongruous association or an adventitious context. The evidence of antiquity, so far as Abbott was now concerned, "is and must be the same here as in Europe, and only when we find the geological and archaeological conditions in accord, i.e. rude implements in undisturbed deposits, can we assert that such evidence has been found" (Abbott 1890:10).

Abbott was going to stand by Trenton on the strength of its geological and archaeological context, not just the resemblance of its artifacts to European Paleoliths. The challenge was clear: if Holmes aimed to reject Trenton as a Paleolithic site, he would have to do so on those same grounds. He would have to go to Trenton.

It was a bold challenge and response, but it was not for Holmes alone. In the event Piney Branch had raised doubts in their minds about Abbott's scientific respectability, he was aiming at the president (Joseph Leidy [1823–91]) and the board of managers of the University's Archaeological Association (his governing board), which included Stewart Culin (1858–1929) and Daniel Brinton. Abbott was on tenterhooks at the museum. Leidy had strongly opposed Abbott's appointment as curator.[9] Culin cared little for him, and Abbott didn't want to give either reason to further their opposition—particularly since the results from his first field season had not gone well. Aiming for greater Paleolithic rewards than had come from the alluvial gravels of the Delaware valley, Abbott had planned "extensive explorations" of the region's caves and rock shelters. If the Delaware Valley was like the Somme, as a stream of distinguished foreign archaeologists had assured him, then surely there must be archaeologically rich caves to match those of Europe.

Abbott selected several promising caves, marshaled start-up funds, then devoted part of the summer to "cave-hunting" (with obvious echoes of Boyd Dawkins). Only, the Delaware proved not to be the Dordogne. Abbott went into several caves, didn't find anything, and quickly exhausted his funds. It was not an auspicious start.

Nevertheless, Abbott sugar-coated it as best he could, assuring the council that negative results were valuable—"principally in experience" (Abbott's)—even if they hadn't produced additions to the museum collection. Besides, there were many more caves with potential. Could more money be made available to continue the work (Abbott 1890:11)? Apparently the council wasn't convinced that contributing to Abbott's experience was a wise use of scarce funds. The following spring (May 1891), this time *before* Abbott went off to the field, they asked him to prepare a detailed proposal, complete with plans, maps, methods, probable results, and costs (Abbott to Council of the University Archaeological Association, May 4, 1891, CCA/PU). Abbott had learned his lesson: rather than continue the risky strategy of excavating possibly unproductive caves, he would go back to a sure thing and "undertake an exhaustive examination, both geological and archaeological," on two islands and over a long stretch of the Delaware valley.

Holmes's Piney Branch work, at the very least, gave Abbott a handy justification to revisit his old ground and "demonstrate beyond all cavil" that Paleolithic remains were found in the Trenton gravel (Abbott to Council of the University Archaeological Association, May 4, 1891, page 3, CCA/PU). Yet despite recognizing the vital role of geology in dispelling doubts about the context and age of the Paleoliths, he still didn't see that excavations, so "impracticable and difficult," were necessary. He would proceed much as he always had, only this time floating downriver by boat, examining otherwise inaccessible cut bank exposures for Paleoliths. Just to insure that a good haul of specimens would come into the museum, he would also look into at least four Indian village sites along the way, put his son Richard in another valley collecting artifacts, and then join Henry Mercer inspecting quarries, excavating mounds, and surveying sites throughout eastern Pennsylvania.[10] He would do it all for under three hundred dollars. The council accepted Abbott's proposal.

In the end, Abbott focused his attention on two islands in the Delaware River (though just how much time he was actually there is unclear). Mercer, who appears to have done the more extensive fieldwork, examined sites in Bucks, Northhampton, and Lehigh Counties. Part of the time they worked together at Mercer's sites, with Abbott himself admitting he played the role of a "mere on-looker" (Abbott 1892c:1, 19; perhaps because of concern that not enough had been done that summer, Mercer had urged Abbott to do more fieldwork that fall, so as to have more material to go into the report [Mercer to Abbott September 11, 1891, CCA/PU]). After Mercer submitted his report on his summer's activities, Abbott could afford on-looking no longer, for Mercer had come onto a jasper quarry north of Limeport, Pennsylvania, and was suddenly sounding a lot like Holmes (Mercer's report is unpublished, but exists in the form of a twenty-five-page letter to Abbott, copied over in Abbott's hand and dated October 27, 1891, Henry C. Mercer Papers, Bucks County Historical Society [HCM/BCHS]).

Included among the flakes and tools at the Limeport quarry were unmistakable turtlebacks of various sizes and in various stages of manufacture, rather like, Mercer admitted, those found at Piney Branch, and for that matter the "Paleolithic" forms from Trenton. But it was just as unmistakable to Mercer that the Limeport quarry had been worked by modern Indians. If these turtlebacks were clearly "blocked-out or unfinished implements," did that mean Holmes was right about the American Paleolithic? Mercer backed off here: after all, Neolithic groups had "but one method of implement manufacture," and naturally along the way it produced typical Paleolithic forms; just as, Mercer supposed, some Paleoliths were themselves unfinished forms. But it was significant, Mercer thought, that none of the purported Paleoliths found to date were made of jasper—only argillite (Mercer to Abbott, October 21, 1891, pp. 85–87, HCM/BCHS).

All this was getting a bit too close for comfort for Abbott, for Mercer's report had even stirred doubts in longtime American Paleolithic advocate

Thomas Wilson (Wilson admitted Mercer's investigations "may decide in favor of paleolithic man or it may not. I stand quite ready to drop all former opinions and yield my assent to such ideas as come from newly discovered facts" [Wilson to Mercer, January 2, 1892, HCM/BCHS]). As Mercer's report was to be sent to the Archaeological Association's Council, Abbott carefully annotated the text, highlighting the absence of *jasper* Paleoliths and insisting that none of the Paleoliths were actually unfinished. The modern Indians might have a use for rudely chipped forms, but "we are not sure how specialized a form was sometimes fashioned by paleolithic man" (Abbott notes in Mercer to Abbott, October 21, 1891, pp. 86–87, HCM/BCHS).

He was being a bit disingenuous here: Edward Paschall, a onetime journalist and an avid collector in southeast Pennsylvania, had found at the Point Pleasant argillite quarry (and sent to Abbott) several rude, "Paleolithic" forms—which were found right alongside and formed a series with more finished artifacts. There was no doubt Point Pleasant was a recently occupied Indian village, and Paschall warned Abbott that he needed to pay more attention to these unfinished argillite forms (Paschall to Abbott, April 5, 10, 24, 1890, UPM).[11]

Yet, rather than have to dance around all of that in his official published report of the summer's fieldwork—which otherwise included long verbatim passages of much of what Mercer had said about the other sites and quarries he visited—Abbott simply gave the Limeport quarry a passing mention. Not a word of Mercer's discussion of turtlebacks and Paleoliths was mentioned—at least in print (Abbott 1892c:28). But in private Abbott snarled at Mercer: "You seem to think that I have been or could be mistaken in these matters. *I Can't*" (Abbott to Mercer, December 1, 1891, HCM/BCHS, emphasis and relative type size as in the original).

Earlier that year, Abbott and Paschall had been haggling over the genuineness of the Lenape Stone, and Paschall—who was himself skeptical about Abbott's Paleoliths—expressed the opinion that it was the responsibility of those who believed it was genuine to prove it. Abbott didn't think so: "I don't quite see it in this light. Why should not 'doubters' *prove* the falsity, if such there is?" (Paschall to Abbott, July 5, 1891, CCA/PU). The doubters, Holmes chief among them, would be happy to oblige.

Yet, it was Mercer who took the lead in planning the following summer's work, outlining—to Pepper, not Abbott—an ambitious plan for fieldwork in eastern Pennsylvania and beyond (Mercer to Pepper, January 13, 1892, UPM). Pepper gave his approval and telegraphed Abbott to come into the museum for a meeting of the 'Committee on the American field' to "lay out a large plan of aggressive work" (Pepper to Mercer, January 14, 1892, HCM/BCHS). But while the plans were evidently well laid out, Abbott's participation was proving to be something less than enthusiastic, perhaps in part because of disagreements about the goals of the work, and the fact that Mercer, ostensibly his subordinate, was clearly codirector of the operation, and had the president's ear. In mid-June of that year, an increasingly frustrated Pepper tried to jump-start matters with a letter to Mercer:

> Have you started on the Excursion yet?
> I earnestly hope you will permit no misunderstanding to exist. Dr. Abbott has written letters from which one would infer that he had nothing to do with the Excursion except to be occasionally with you. You know what the action of the Board was. I hope you will keep him up to the mark. I am deeply interested in his success and am anxious that he should do, with you, this summer, a good piece of scientific exploration, bring in a scientific report and let us have an interesting scientific exhibit of the objects [Pepper to Mercer, June 15, 1892, HCM/BCHS].

Pepper used the term "scientific" not once but three times (and did so again in a letter the following month [Pepper to Mercer, July 10, 1892, HCM/BCHS]). Obviously, he was very concerned about the scientific legitimacy of the operation—nominally under Abbott's direction—and made those concerns explicit to Mercer: "You know that Dr. Abbott's contract with the Museum will expire in November of this year [1892], and I want to have it renewed on terms advantageous to himself. It is very clear to me that he has not been gaining in the confidence of our colleagues as a thoroughly scientific archaeologist. I appreciate his qualities so highly myself that I want him to show more scientific method in his work, so as to justify us in entrusting important explorations to his care" (Pepper to Mercer, June 15, 1892, HCM/BCHS).

But the fieldwork was not going well, either in terms of the recovery of material ("I am sorry you drew some blanks at first" [Pepper to Mercer, June 26, 1892, HCM/BCHS]), but also in terms of Abbott's establishing himself either in the field or in his relations with his board members:

> [Dr. Abbott] must not let himself get irritated and disturbed just because other people who are giving their money and devoting their

> thoughts to the subject have their ideas as well as he has. As you very well know, I am utterly ignorant of the whole subject . . . [but] I believe in it heartily, and I want to see our Department of Archaeology grow; but I see at once that there are so many different views and so many different ways of looking at it that it needs a great deal of good humor, patience, and co-operation all around [Pepper to Mercer, June 26, 1892, HCM/BCHS].

Pepper didn't have to speak the obvious. These were not qualities Abbott possessed.

In the end, Mercer was alone in the field most of that summer, duly reporting his activities to Pepper—who by mid-July was no longer feigning any pretense that Abbott was playing any part at all in the fieldwork. In fact, Pepper was wondering just what his salaried curator was up to. So far as he could tell, Abbott was not engaging in any museum work and apparently wasn't even bothering to come into Philadelphia: "As he has been receiving a salary at the rate of $100 per month there certainly should be some explicit understanding with regard to his official duties" (Pepper to Mercer, July 13, 1892, HCM/BCHS). It was becoming a matter causing Pepper "much anxiety."

Abbott was not much help to himself, either, for when he did come into the museum, many of his days were apparently spent doing nothing at all (Hinsley 1985). Abbott evidently was not cut out for museum work—but then, it appeared he was not cut out for any form of employment, however much he needed the income. (A decade later, Abbott found himself back in a museum for a visit, and it seemed to him "strange and dream-like." As he mused that day: "Out-door work in archaeology is beyond comparison, the most delightful of occupations, but the museum and contact with humanity—I shudder at the thought" [Abbott Diary, February 24, 1903, CCA/PU].)

But then there was incentive for Abbott to stick close to Trenton. That summer Holmes, along with WJ McGee (1853–1912)—a onetime supporter of the American Paleolithic (e.g., McGee 1888) but now increasingly strident critic—spent time in Trenton examining the archaeology. At times, they were accompanied by Abbott as well as Ernst Volk (1845–1919), whom Putnam had hired to replace Abbott as the Peabody Museum's collector-on-the-ground at Trenton. Also joining the field party was glacial geologist Rollin Salisbury (1858–1922) of the University of Chicago and United States Geological Survey. He had been "borrowed" by the New Jersey state geologist to map the Pleistocene formations of the state and by the summer of 1892 had worked his way to the Delaware valley near Trenton. The gravels at Trenton were, in Salisbury's view, part of a valley train of debris deposited by outflow streams from the glacial front. But whether the specific, Paleolithic-yielding gravels at Trenton were contemporaneous with the ice front when it stood ~50 miles north of Trenton (Wright's view) or had been reworked long after the glacial period ended was another matter. The fact the gravels were not continuous up the valley (contra Wright) suggested the latter to Salisbury, or at the very least the necessity of carefully determining the age of the formation in which any artifacts were found (Salisbury 1893:106, 113–14; cf. Wright 1892:244–45).

As a result, by the time Holmes boarded the train for Trenton, Salisbury was already quite familiar with the regional glacial geology, and Holmes was at pains to insure Salisbury would be there to jointly examine the Trenton locality. Holmes wouldn't have to rely on Lewis's or Wright's work from the previous decade. While at Trenton, Holmes and company examined the walls of a long, wide, and deep sewer trench through the gravels, which had just been dug by the city: but search as they might, no Paleoliths were found. To confirm their absence, Holmes subsequently arranged for William Dinwiddie, a bureau assistant, to spend "upwards of a month" "in the trench with the workmen," scouring the gravels for any sign of artifacts. He didn't find any either (Holmes to Salisbury, July 2 and 12, 1892, Rollin D. Salisbury Papers, University of Chicago [RDS/UC]). It all confirmed to Holmes that Abbott's claims were utterly unconvincing, and he pronounced as much at the AAAS meetings in Rochester a few weeks later.

After Trenton, McGee and Holmes headed up the Delaware valley to examine the Point Pleasant argillite quarries, where they saw just what Paschall and Mercer had seen: "an Indian workshop with abundant material . . . including several specimens corresponding to the typical turtle-backs hitherto supposed to occur [only] in situ in the Trenton gravel" (McGee to Salisbury, July 26, 1892, William J. McGee Papers, Library of Congress; Holmes to Salisbury, July 26, 1892, RDS/UC). Six months later, in his detailed critique of Abbott's Trenton gravels, Holmes would gleefully report that upon seeing the Point Pleasant sites Abbott was "entirely at a loss to explain the occurrence," though the explanation seemed to Holmes to be "at once apparent to any one not utterly blinded by the prevailing misconceptions" (Holmes 1893b:34).

Just a few days after Holmes' party left Trenton, George Frederick Wright and others arrived to spend the day, escorted about by Abbott and Volk. Abbott, whose diary entry about the visit from Holmes's group was uncharacteristically bland, was downright effusive when it came to recording the visit by Wright's party. Despite temperatures over 100 degrees Fahrenheit, Abbott wrote it was "a most delightful day" (Abbott Diary, July 23–24, 29, 1892, CCA/PU).

He needed a few of those, for things were not going well. Shortly after his visitors left Trenton, Abbott confided in his diary that he felt his curatorial position was "slipping away." For that he blamed Daniel Brinton and Culin, who were, he thought, "both liars"—but who, unfortunately for Abbott, were also both on his governing board (Abbott Diary, August 3, 1892, CCA/PU). What he might have heard is unknown, but he knew he would have to be on his best behavior, and his Trenton gravels would have to withstand critical scrutiny. No easy tasks, those. Especially not with the very public cannonade of the American Paleolithic that began just a few weeks later at the Rochester AAAS meetings and then escalated through the fall and winter of 1892–93.

The Great Paleolithic War

The Great Paleolithic War, as the participants came to call it, was sparked by a rapid fusillade of vicious reviews, including several by Salisbury and WJ McGee, of Wright's *Man and the Glacial Period* (which appeared in the fall of 1892). The war was fueled by a resentment of the arrogance and heavy-handedness of these government scientists and played out in a series of increasingly angry exchanges between Paleolithic proponents and opponents (both archaeological and geological) (Meltzer 1983, 1991, 1994). Abbott had counseled Wright to ignore his critics (Abbott to Wright, February 7, 1893, GFW/OCA), but when the attack turned on Trenton, as it inevitably did, it was advice he was utterly unable to follow himself.

Abbott, in fact, had tried to preempt the attack, or at least fortify the home front. Leaving Philadelphia one afternoon that fall, he'd run into Brinton, and they'd both returned to the museum and talked for over an hour about "gravel beds and antiquity of man" (Abbott Diary, November 2, 1892, CCA/PU). But if he had hoped to swing Brinton to his side, he was too late. Brinton had already written his weekly *Science* column ("Current Notes on Anthropology"), which two days later publicly declared what everyone privately already knew:

> For two or three years past there has been in the air—I mean the air which archaeologists breathe—a low but menacing sound, threatening some dear theories and tall structures, built, if not on sand, at least on gravels offering a scarcely more secure foundation. . . . the alarming discovery has been made that a great many of what we have heretofore called "paleolithic implements" display with fatal clearness the peculiar earmarks of these "quarry rejects," hinting, therefore, that they were never really implements at all. . . . it has even been hinted that the "Trenton gravels" of our own land, may have to lose their laurels in the light of this discovery [Brinton 1892b:260–61].

Putnam had heard the same menacing sound and had started to get nervous. He wrote Abbott urging him to "do all you can" to show the geological context and character of the Trenton gravel that yielded the Paleolithic specimens (Putnam to Abbott, November 14, 1892, PMP/HU).

Abbott could hardly stand idly by, even if it meant publicly contradicting Brinton, one of the members of his governing board. He shot back in *Science* a week later at "the geologists at Washington; and those [like Brinton, though he was not named] that look upon them as little gods." If these critics were correct, he snarled, why were only the rude argillite forms found in gravel deposits, and not any other artifacts of the Indians? If these were quarry rejects, where were the "chips resulting from their fashioning?" And why were signs of use critical to the issue? Abbott asked, "does the spear or arrow point show signs of use?" Most insulting of all, who were these self-styled "expert geologists" to lecture him about gravel deposits, or judge the age of the Trenton gravels, when they cannot agree amongst themselves? Abbott simply couldn't imagine how anyone who had not spent as much time as he walking the gravel banks of the Delaware could assert greater knowledge of the geological situation than he possessed. After all, Abbott declared:

> I lay a claim to a smattering of gravel-ology. I have lived on pebbles so long that I have become flinty-hearted so far as criticism is concerned, and when I find gravel stratified and unstratified I know and assert the difference; and when a paleolithic implement is found in gravel beneath layers of sand and pebbles, beneath huge bowlders . . . no reasonable person should want another to tell him that the two were laid down together. . . . [But then] Up pops some "authority" and de-

> claims the possibility that the ground was washed from beneath the big stone and the implement slipped in. Well, we can go on supposing till the crack o'doom, but as to proof, that is another matter. These geological jugglers will prove yet that the Indians bought the Delaware Valley from William Penn [Abbott 1892a:271].

Perhaps. But as Holmes (one of the "geological jugglers"—along with McGee) argued two weeks later in *Science,* demonstrating that artifacts were truly found in glacial gravels (leaving aside the issue of whether those artifacts were even finished tools, let alone finished Paleolithic tools) required the work of "competent and reputable observers of geological phenomena" (Holmes 1892:296). That Holmes believed no glacial-aged artifacts had yet been found made it quite clear he did not consider Abbott's claims to knowing some "gravel-ology" to be tantamount to competence.

A testy Abbott penned his reply the very afternoon Holmes's paper arrived, and in a gambit guaranteed to offend the humorless Holmes, Abbott put the matter in verse (Abbott 1892b:344–45):

> The stone are inspected,
> And Holmes cries "rejected,
> They're nothing but Indian chips"
> He glanced at the ground,
> Truth, fancied he found,
> And homeward to Washington skips.
>
> II.
> They got there by chance
> He saw at a glance
> And turned up his nose at the series;
> "They've no other history,
> I've solved the whole mystery,
> And to argue the point only wearies."
>
> III.
> But the gravel *is* old,
> At least so I'm told;
> "Halt, halt!" cries out W. J. [McGee],

"It may be very recent,
And it isn't quite decent,
For me not to have my own way.

IV.
So dear W. J.
There is no more to say,
Because you will never agree
That anything's truth
But what issues, forsooth,
From Holmes or the brain of McGee.

Holmes was not amused. Not by this, and not by Abbott's just published report on his 1891 University Museum-sponsored Delaware valley work, in which he vigorously defended the archaeological and geological integrity of his Trenton sequence against the "disgraceful articles in pretentious periodicals, written by persons wholly ignorant of the subject" (Abbott 1892c:2). Whether Holmes saw himself in that caricature (and whether Abbott intended it for Holmes alone, or Holmes and McGee) is unclear, but its consequences were evident enough. Holmes had intended to review the report for the *Journal of Geology* but on reading it had second thoughts: "It contains such vile personal insinuations and such a mass of errors, geologic, archaeologic, and otherwise, that it cannot be mentioned save to be damned. I may perhaps write something for 'Science,' which is the trough into which the slops are thrown" (Holmes to Salisbury, December 31, 1892, RDS/UC).

True to his word, Holmes replied in *Science* a few weeks later (January 20, 1893). He avoided any mention of Abbott by name. It didn't matter, however, since it was obvious he had Abbott in mind (virtually the same language was used in Holmes's long-awaited critique of the Trenton gravels, which appeared almost simultaneously). Holmes blamed the American Paleolithic on the blunders and misconceptions of "amateurs" who worked and wrote for themselves, who had little scientific understanding of stone tool making, let alone of geological age and context, and who couldn't see the critical distinction between evidence of a glacial human presence (age) and evidence for humans in a Paleolithic stage (evolutionary grade) in America. The American Paleolithic, Holmes insisted, was so "unsatisfactory and in such a state of utter chaos that the investigation must practically begin anew" (Holmes 1893c).

If Abbott didn't get the message then, Holmes spelled it out all too clearly in a lengthy critique published in the *Journal of Geology*, which asked the decidedly nonrhetorical question, "Are there traces of man in the Trenton gravels?" (Over the first few months of 1893, Holmes also published critical examinations of the Paleolithic claims from the Tidewater region; the Loveland, Madisonville, and Newcomerstown, Ohio, finds; and the Little Falls, Minnesota, Paleoliths, in addition to Trenton [Holmes 1893a, 1893b, 1893d, 1893f].)

At Trenton, as with each purported Paleolithic case, Holmes came away convinced there were fatal flaws. The alleged Paleolithic specimens were all "typical rejects of the modern blade maker," mistakenly identified as Paleolithic. None were in true glacial-age deposits: those artifacts found in situ in "gravels" were actually in geological strata redeposited in postglacial times. And here Holmes turned Carville Lewis against Abbott, claiming Lewis and Leidy both thought the Trenton gravels were reworked. Abbott sputtered angrily into his diary: "Leidy never visited the Trenton gravels but once . . . [and] Carville Lewis instructed by me, not vice versa" (Holmes 1893b:24; Abbott Diary, undated but ca. January 1893). No matter, Holmes baldly stated that any and all of the specimens must have fallen into otherwise intact Pleistocene deposits, while others had been collected from redeposited gravels or other, recent formations—such as the talus slopes that flank the Delaware valley. But none had been recognized as being in secondary deposits and contexts, which was no surprise, so far as he was concerned: "Talus deposits form exceedingly treacherous records for the would be chronologist. They are the reef upon which more than one paleolithic adventurer has been wrecked" (Holmes 1893b:27). To show how easily artifacts on the surface could tumble into deeper deposits, Holmes provided hypothetical illustrations of massive slope waste at Trenton.

That none of the Paleolithic claims could be or had been found in place and verified by independent, competent observers only confirmed Holmes's suspicions that the reason he never saw any artifacts in place in Pleistocene deposits in those sites when he visited—as, for example, in the deep sewer trench at Trenton—is that there were *no* artifacts in place in Pleistocene deposits in those sites. They "are and always were wholly barren of art [artifacts]" (Holmes 1893b:21, 23).

As for Abbott's elaborate (and allegedly stratified) sequence at Trenton of Paleolithic/Eskimo/Indian archaeological cultures, Holmes suggested (in an obvious play on Abbott's use of the same allusion) that, had "William Penn paused [at intervals] in his arduous traffic with the tawny Delawares," he would have seen at different moments over a brief period of time an "uncouth savage" making a stone tool from which he could have gleaned the story of the ages—or at least the stone tool evolution (phylogeny) as it played out in the process of stone tool manufacture. But since he didn't, "Two hundred years of aboriginal misfortune and Quaker inattention and neglect have resulted in so mixing up the simple evidence of a day's work, that it has taken twenty-five years to collect the scattered fragments, separate and classify them, and to assign them to theoretic places in a scheme of cultural evolution that spans ten thousand years" (Holmes 1893b:34). Abbott was the inattentive Quaker in question.

Clearly, there was in Holmes's mind neither archaeological nor geological evidence that the purported American Paleoliths were anything but the remains of the historic Indian (Holmes 1893b:18, 27). As Trenton made clear, "The evidence upon which *paleolithic man* in America depends is so intangible that, unsupported by supposed analogies with European conditions and phenomena, and by the suggestions of an ideal scheme of culture progress, it would vanish in thin air; and if the theory of a *glacial man* can summon to its aid no better testimony than that furnished by the examples examined . . . the whole scheme, so elaborately mounted and so confidently proclaimed, is in imminent danger of early collapse" (Holmes 1893b:37, emphasis in original).

The American Paleolithic and the claims for the Trenton gravels were no more than the flawed outcome of work done by "enthusiastic novices with fertile brains and ready pens," who were "utterly blinded" by their misconceptions and unable to overcome their "unscientific methods and misleading hypotheses" (Holmes 1893b:29, 34–35).

Tough words, those. Holmes thought his Trenton critique "the best work I have ever done" (Holmes to Salisbury, December 8, 1892, RDS/UC). Abbott tried to laugh it off, with yet another response and another round of doggerel (Abbott 1893). But there had been heavy damage inflicted—Abbott's governing board at the University Museum had been watching closely—and it ultimately proved fatal.

War and Its Aftermath

The dispute over the Paleolithic, nominally about a straightforward archaeological issue, was much more than that—as I have noted previously (Meltzer

1991:26–27, 1994). Most prominently, it was about change within the archaeological community, and the very self-conscious and potently polemical efforts to establish an exclusive and professional class of archaeologists. Abbott was a different sort of archaeologist. Like Putnam, he'd come out of a natural history intellectual tradition, but in some ways Abbott had never left it.

At the same time as he was writing of an American Paleolithic, Abbott was producing dozens of papers in zoology and natural history, on a range of topics from the habits of crayfish to winged ants, the intelligence of fish, feeding habits of kingfishers, the nests and eggs of the thistle-bird, color-sense in fishes, and animal weather lore (Anonymous 1887:550–51). Odd as that juxtaposition might appear, in a sense his archaeology and natural history were but two sides of the same whole. But not in ways that might be expected.

Abbott was, of course, well aware of current intellectual developments in the sciences and occasionally made mention of how his observations might speak to an aspect of evolutionary theory, but otherwise his work was largely untouched by those larger currents. In fact, it was not within any particular theoretical framework at all. Rather, Abbott was first and foremost a devout observer of the natural world (including human artifacts), or at least his small corner of it along the Delaware valley. Abbott's vision rarely extended beyond this area—though his ambitions would.

Abbott's writings about the place hearken back to an earlier time and tradition in natural history—that of the gentlemanly naturalist who observed and cataloged pieces of nature. The endgame of his natural history was often a lyrical, firsthand description that illustrated nature's charms and intricacies, all the better if it involved a species new to science.

The endgame of his archaeology was little different: a description of artifacts and their chronological place. The odd artifacts he found—whether from the drift or elsewhere—were no different than the odd behavior he observed among birds or fish. They were all objects to be cataloged and described. Understanding how those pieces might fit into a larger scheme of human or natural history was neither Abbott's interest nor concern. His theoretical framework, insofar as one existed, was an expedient one, in which certain principles (the "theory of the gradual progress of mankind") were invoked to support or provide a rationale for his observations. But largely observations were made and the material arrayed chronologically and historically within a commonsense framework, which aimed to accommodate his facts within the current European scenario about human prehistory, in the absence of any broad theory. Proof in such instances was self-evident.

As a consequence, Abbott found himself being left behind by the tide—though he never appreciated just how badly behind he was. Holmes, McGee, and other Paleolithic critics had research programs, theoretically derived and methodologically coherent. And they used barely disguised code to elevate their work (that of "scientific men" in this "new era" of archaeology) and disparage the work of the "old archaeology" (Holmes 1893e:135) practiced by those like Abbott, whose "blunders and misconceptions" were not unexpected, given that "the field has, up to this time, been occupied mainly by amateurs who have not mastered the necessary fundamental branches of science."

Without denying the polemical aims and language of the critics, it is indeed the case that Abbott's efforts were widely perceived as amateurish and not altogether scientific, even by supporters (like Pepper) and by allies like Putnam and Wright. Worse, there had long been lingering doubts about Abbott's veracity and reliability—likewise among his Paleolithic compatriots: Henry Haynes blamed the difficulty of establishing a Paleolithic on "doubts in regard to Abbott's truthfulness" (Haynes to Wright, November 16, 1892, GFW/OCA). Putnam too had misgivings:

> This is sub rosa, but you know that every time Abbott is the discoverer [of an artifact in ancient deposits] it will be said that it is strange that no one else finds them. So I think, in the future diggings we must, without injuring his sensibilities, be very careful who is using the trowel and first finds a specimen. I think Dr. Abbott should take the same position I do, and let others have a chance to judge for themselves rather than from his discoveries. But you know what a peculiarly sensitive man he is, so that I must ask you not to lisp it that I might make this suggestion [Putnam to Wright September 7, 1897, GFW/OCA].

Abbott was well aware such things were being said about him, telling Wright on one occasion, and strictly tongue-in-cheek, "Come [to Trenton] by all means, and I *won't* salt the ground in advance." Abbott could joke, but such doubts hardly helped his reputation or his cause.

Archaeology was coming into itself as a discipline. The previously hazy boundary separating amateur and professional was coming into focus, and it was apparent to all which side of the line Abbott was on. Soon Abbott would lose the only job in archaeology he'd ever have.

It was, perhaps, inevitable. Placed in a position at the University of Pennsylvania he had so long coveted, he discovered the reality of museum work and the demands of aggressive fieldwork were not nearly so appealing as imagined from outside. He never quite recovered from his dismal start in the summer of 1890.

Yet, Abbott's undoing wasn't solely the result of poor job performance. The Great Paleolithic War had taken its toll. Like Putnam nearly two decades earlier, Culin was concerned that Abbott's "loose methods" were detrimental to the reputation of the University Museum and undeserving of receiving the "official approval of respectable institutions" (Culin to Holmes, August 1, 1892, William H. Holmes Papers, Smithsonian Institution Archives [WHH/SIA]). It was clear by the late summer of 1892 that Culin did not anticipate Abbott would be at the University Museum for the long term, but Mercer might (Culin to Holmes, August 1, 1892, WHH/SIA). Since Culin and Brinton, on Abbott's board of supervisors, were in a position to do something about the matter, and Pepper was growing increasingly dismayed at Abbott (and thus not likely to be protective of him), the outcome was inevitable.

Abbott had sensed it too and by the late summer of 1892 was writing Putnam to tell him that he would have to "give up" when his museum year ended that fall. Abbott, of course, took none of the blame. "What I was warned, but laughed at, three years ago, has more than come to pass, and I cannot see how with a trace of self-respect, I can remain, for I do not believe the offensive actions of the Board of Managers will be re-called" (Abbott to Putnam, August 28, 1892, FWP/HU). Apparently Abbott was offended that the board recognized it was Mercer who deserved the credit for the fieldwork done under Abbott's auspices; he was unhappy about not receiving a two-hundred-dollar annual raise; and he chafed at being subordinate to Culin (Madeira 1964:20). Once again, Abbott made overtures to the ever-patient Putnam to take him back. Putnam, fearful of being seen as meddling in another institution's affairs and happy with the arrangements he'd made with Ernst Volk (and perhaps only too happy not to have to deal with Abbott as an employee), took several months to answer Abbott's letter, and then urged Abbott to stick it out at the University Museum—that he could offer nothing at the Peabody. He warned Abbott, too: "If you make another change it would injure you very much" (Abbott to Putnam, August 28, 1892; Putnam to Abbott, November 14, 1892, FWP/HU).

Abbott hung on through his annual term, managed to get reappointed, but spent much of 1893 embroiled in the Paleolithic controversy, which—unfortunately for him—both Culin and Brinton were closely following. Worse, they each considered Holmes's Trenton critique (as Brinton put it) "a very strong case against the Abbott finds," one that would "greatly aid in placing the Delaware Valley relics in their proper cases" (Brinton to Holmes, March 9, 1893, WHH/SIA). To which Culin added, the matter was now perfectly clear and convincing save to those (he had Abbott in mind) "who court notoriety at any expense" (Culin to Holmes, March 29, 1893, WHH/SIA). There was little Abbott could do to salvage his standing at the museum and through the summer of 1893 little he actually tried to do—fieldwork included (though he did visit Mercer in the field on occasion).

Finally, on October 13, 1893, Abbott got official word from Brinton that his long-faltering curatorship was over. Henry Mercer was the heir apparent. Two weeks after he was fired, Abbott had a long, angry meeting with Mercer and Pepper, Abbott yelling he'd been "swindled" by the museum and Pepper—who must have felt he'd already stood by Abbott too long—swearing he would not make any financial settlement with Abbott. Mercer stepped in to calm matters. Abbott, fearful he'd lost any chance of receiving severance pay, offered to recall what he'd said. Pepper graciously accepted the offer, agreed to a settlement, and Abbott signed a receipt to that effect. So ended Abbott's connection with the museum, a connection that had been "a veritable thorn in my flesh for over two years." Badly shaken, Abbott went off to a nearby tavern with Mercer, where he pressed him to accept the curatorship as an honorary position. Mercer did (Abbott Diary, October 13, November 1, 1893, CCA/PU).

A month later, the University Museum sent Abbott an invitation to attend a lecture by WJ McGee, of all people. Abbott was not amused by the irony of it all: "What a pack of asses these people are!" (Abbott Diary, December 11, 1893, CCA/PU).

Acknowledgments. Archival research reported in this paper was supported by a grant from the National Science Foundation (DIR-8911249). The writing was facilitated by a Southern Methodist University Faculty Research Fellowship Leave, for which I am most grateful. I thank Stephen Williams for comments on a draft of this paper.

Notes

1. Abbott received his M.D. from the University of Pennsylvania in 1865. His course of study there involved anatomy classes with Joseph Leidy, who encouraged Abbott's interests in natural science (Aiello 1967:210). Abbott's own view of his career path emerges in the "Sketch of Charles C. Abbott" that appeared in the *Popular Science Monthly* in 1887. The author of the "Sketch" was anonymous, but by its content and style I infer it was written by Abbott himself or at least someone well coached by him.

2. Through the 1870s, Abbott corresponded with Lubbock about artifact classification and function, in addition to sending him artifacts (e.g., Abbott Diary, April 13, 1875, February 10, 1876, CCA/PU; also Abbott 1872:199, 205, 221).

3. Wyman, the first curator at the Peabody Museum, had by then acquired several large collections, including the private collection of French prehistorian Gabriel de Mortillet. In it were European Paleolithic specimens, including artifacts from St. Acheul, Abbeville, and Amiens, "the scene of the original investigations of Boucher de Perthes" (Wyman 1869:6). Wyman was keen to have these in America for the purpose of "making direct comparison between the implements of the stone age of the old world and the new" (Wyman 1869:11), as he evidently did in this instance.

4. Croll and Geike's correlation of glaciation with the earth's orbital eccentricity and precession foreshadowed the work of Milutin Milankovith in this century.

5. Abbott moved onto the homestead in 1874 and lived there until November of 1914, when a disastrous fire on Friday the thirteenth destroyed Three Beeches. Today, nearly ninety years later, the property remains undeveloped and the tree-covered stone foundation of Three Beeches is still visible.

6. The sums paid Abbott, Palmer, and others are itemized in the Report of the Treasurer in the tenth, eleventh, and twelfth *Annual Reports of the Peabody Museum.* In the thirteenth and subsequent *Annual Reports* individual payments were no longer itemized.

7. By the time Dawkins published his testimony, a smug Abbott would

belittle it: "I do *not* consider Prof. Dawkins one whit better authority than a half dozen others who have been here" (Abbott Diary, written November 17, 1881, alongside the entry for November 19, 1880, CCA/PU). That dismissal is partly attributable to Abbott's having finally read *Early Man in Britain* and realizing that if Dawkins's scheme was correct (that is, if the "river-drift men" vanished from the face of the earth without any links to living people, while the later "cave-men" of Europe were the ancestors of the Eskimo), then he (Abbott) would be forced to abandon his own. Obviously, Dawkins could not be *that* good an authority (if Dawkins knew of Abbott's efforts to link the "river-drift men" with the Eskimo, he simply ignored them).

8. The skirmish was relatively brief and erupted after crews for Cyrus Thomas's Mound Survey moved into Ohio's Little Miami valley, where Putnam had been working since 1880. Its essence is neatly captured in two letters written in the summer of 1884. The first, from Baird to Thomas (August 4, 1884): "I suspect you are quite correct in your impressions in regard to Putnam. We cannot admit for a moment his right to monopolize any field of archaeological research." The second, from Putnam to Asa Gray (which Gray forwarded to Baird!): "Many thanks for your contribution for the Ohio work, which is going all right at present, notwithstanding the fact that the Bureau of Ethnology of the Smithsonian Institution have sent their men to the valley . . . just above our present operations in the valley. . . . it seems as if there were plenty of places in Ohio where the S.I. could work without destroying part of the evidence I wish to secure" (August 15, 1884). Both letters are in CT/MSP.

9. In a very telegraphic diary entry on May 7, 1885, Abbott refers to patching up the "bitterness of past experiences" with his onetime teacher, Joseph Leidy of the Academy of Natural Sciences. What caused that bitterness is unexplained here, but it apparently occurred fifteen years earlier (1870) and again in 1880 in connection with Dawkins's visit. Whatever the cause, relations with Leidy were never completely mended, hence his opposition to Abbott's appointment as curator (Abbott Diary, May 7, 1885, and December 28, 1889, CCA/PU).

10. Mercer [1856–1930] was described at the time as a collector, traveler, and "man of means." A Harvard graduate, he is best known for having published *The Lenape Stone* (1885), which apparently sparked his interest in archaeology. He joined the university's archaeological association in 1890 and was soon thereafter on the museum's governing board (Dyke 1989:45; also Williams 1991:117–21).

11. Mercer himself was looking for argillite quarries near Trenton that winter (Mercer to Abbott, December 9, 1891, CCA/PU) and would a few years later seize on the evidence from the Point Pleasant [Gaddis Run] quarries to make the same point as Paschall, and also use it to cast doubt on the veracity of the Trenton Paleoliths (below, and Mercer 1897:34ff, and also Abbott 1897).

5

Restoring Authenticity

Judging Frank Hamilton Cushing's Veracity

David R. Wilcox

When Frank Hamilton Cushing died in April 1900, many of his memorialists (*American Anthropologist* 1900) remembered with awe his abilities to fabricate Indian crafts so correctly that they appeared to be as authentic as their Indian-made counterparts. Indeed, examples of Zuñi war gods he restored or may have made are regarded as authentic by the Zuñi, who want them repatriated (TJ Ferguson, personal communication, 2000); he was, after all, an initiated bow priest of the Zuñi, who had access to firsthand information about these objects.[1] During the early 1890s, Cushing was of special value to the National Museum because he was able both to produce "authentic" costume elements and to arrange them on mannequins in comparative groups at the Columbian World Exposition in 1893 (Cushing 1893, Wilcox 1999) and to install the Smithsonian's Pueblo Hall the following year (*Washington Post* 1894). Through these manual skills and his literary abilities Cushing sought to achieve a kind of poetic brotherhood with Indian people.[2]

But can anyone really do this? Several of Cushing's contemporaries thought not and attacked his veracity, questioning both the authenticity of several aboriginal artifacts he claimed to have found and the possibility of the kind of poetic knowledge he sought.[3] Even before his death, archaeology as a profession was developing along very different lines than those Cushing advocated.

Inventing Archaeology as a Profession

What most fascinates us today about Cushing is the contemplation of what might have been had his approach to ethnological archaeology prospered and evolved. In 1917 Clark Wissler heralded the appearance of a "new archaeology" based on the study of time relations. The profession of the "old archaeology" crystallized in the early 1890s in connection with the great public success of Americanist studies promulgated by Frederic Ward Putnam's plans for the Columbian Exposition (Dexter 1966).[4] By sending some fifty-five researchers into the field, he doubled or tripled the number of practicing archaeologists (see Powell 1890), and the resulting Field Columbian Museum in Chicago was paralleled by the rise of large museums in other American cities that began to compete for the treasures of world culture just as America reached its first flood tide of economic hegemony (Wilkins 1970, Kennedy 1987). No longer would Americans have to make do with casts as they did in the 1870s (Harris 1962); now they set the pace in the world market for originals of art and artifacts (Strouse 1999).

Jobs for archaeological curators were thus created, and a mad rush to collect, collect, collect ensued (Cole 1985). This was the period in art and architecture called the American Renaissance (Wilson 1979, Hughes 1997), when neoclassical values of civic pride were funded by the emergent economic elite, who built museums to bring enlightenment to the masses and to educate them to America's new role in the world. Even before World War I an intellectual and cultural reaction had set in, led in part by the New York avant-garde (May 1964). It was in that context that the alliance between Elsie Clews Parsons (Deacon 1997), a child of the elite and an exponent of the avant-garde, and Franz Boas created the movement known as modern anthropology. Wissler gave jobs to many of its most promising luminaries and contributed his own guidance on research directions (Fowler and Wilcox 1999, Snead 2001).

Before there was an archaeological profession that could be redirected, Frank Hamilton Cushing had a vision of what archaeology could be. Before the Columbian Exposition there were some institutional structures, such as the Peabody Museum, the Smithsonian Institution, and the Bureau of Ethnology, but they were guided with patriarchal and familial values by men like Spencer Baird and John Wesley Powell, who provided a support system for brilliant young men like Cushing, who lacked college degrees. There were few means of professional certification and little disciplinary culture that specified standards of acceptable research protocols.[5]

After four years at Zuñi (Green 1979, 1990), Cushing had an idea about how to apply his ethnological knowledge to archaeology, and in 1886 Mary Hemenway of Boston agreed to give him that chance (Hinsley and Wilcox 1995).[6] In a forty-five-page typed letter to her from Camp Hemenway in the

Salt River valley, Arizona Territory, in August 1887, Cushing systematically set forth his conception of what the Pueblo Museum they planned to build should be in its organization, exhibitions, field projects, and humanitarian goals (August 2, 1887, HFL/CLB 2:169–212). Among other things, Cushing proposed to bring young men into the Southwest where they would first learn ethnology by participant observation and then learn to apply that knowledge, as he had, to the interpretation of the archaeological record.

Before leaving Mrs. Hemenway's summer residence at Manchester-by-the-Sea the previous fall, he had written out a research design for his work in the Southwest.[7] A primary reason for Cushing's failure to realize his vision was that his patroness had much more modest goals for the expedition, involving completing his studies of Zuñi origin myths and preparing a Zuñi-English dictionary.[8] She thus set aside his research design,[9] and her chairman of the board of the expedition, William Torrey Harris, tried to explain to Cushing in a San Francisco hotel room in July 1888 just what was expected of him.[10] Cushing persisted, however, to do what he thought best, hoping his eloquence would carry him through. He used Mrs. Hemenway's funding to enlarge his house at Zuñi as a field station (fig. 5.1) and to hire a group of young men whom he began to train as ethnological archaeologists.[11] The Hemenways later repossessed the house to recover some of their costs for the expedition.[12] Recalled to Boston in October 1888, his control of the expedition began to unravel, and by July 1889 he was replaced as director by Hemenway board member Jesse Walter Fewkes.[13]

What clinched Cushing's loss of command were the accusations of his field secretary and future brother-in-law, Frederick Webb Hodge, who claimed he owed thousands of dollars to the Hemenways.[14] What turned Hodge from a loyal employee to a lifelong denigrator of Cushing's abilities and vision (see Hodge 1945)? Cushing thought it was jealousy, and there is evidence to support this view.[15] Although he was in charge of the field accounts, after there was an overdraft at a New Mexico bank, Hodge went on a jaunt to Mexico with the expedition's photographer, E. H. Husher (Hinsley and Wilcox 1995), and was in no condition upon returning to answer the anxious questions Cushing was raising.

At first Hodge suggested that receipts could be manufactured to balance the ledgers (Hodge to Cushing, February 20, 1889, HFL), but when Cushing appointed Charles Garlick as field director,[16] Hodge may have thought he was being made the fall guy, and so he turned 180 degrees and resigned, blowing the whistle on Cushing.[17] The fact that he had become engaged to

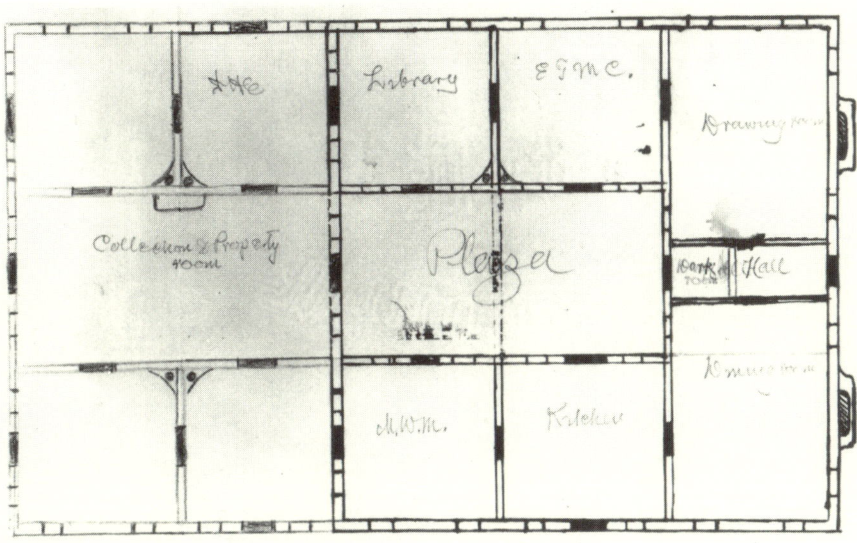

5.1. Frank Cushing's plan for enlarging his Zuñi house. Courtesy of the Brooklyn Museum of Art Archives, Culin Archival Collections, Cushing sketches, Hemenway Expedition (6.3.025 #1041), Brooklyn.

Margaret Magill, but that Cushing, as the senior male in the family, had refused his consent to their marriage because Hodge did not have a permanent job further exasperated[18] Hodge's frustrations with a man whose theories of llamas in Arizona and other interpretations he quietly disdained.[19] From the time Cushing and Emily Magill married in 1882, Emily's younger sister Margaret had traveled with them, but when the Cushings returned east in 1888 Margaret remained at Zuñi—with Fred. Understanding all of this, but helpless in the Garfield Hospital, Cushing drew a prescient symbol of his dilemma, a Gordian knot (SWM, MS 6 HAE 1.18).

That was only the beginning. Slowly recovering his health, Cushing withdrew to the house of his friend in western New York, the banker Ezra Coann.[20] In August 1891 Emily submitted all their receipts and financial records for the expedition to Coann for an analysis. He found not a dollar out of place.[21] Soon, however, Cushing heard that Hodge and Fewkes were spreading new calumnies—that he was a drunkard and had faked numerous artifacts, including the most prized one, an inlaid turquoise toad (fig. 5.2).[22]

Meanwhile, however, John Wesley Powell reinstated Cushing as an assistant ethnologist at the Bureau of Ethnology, where Hodge, too, now worked,

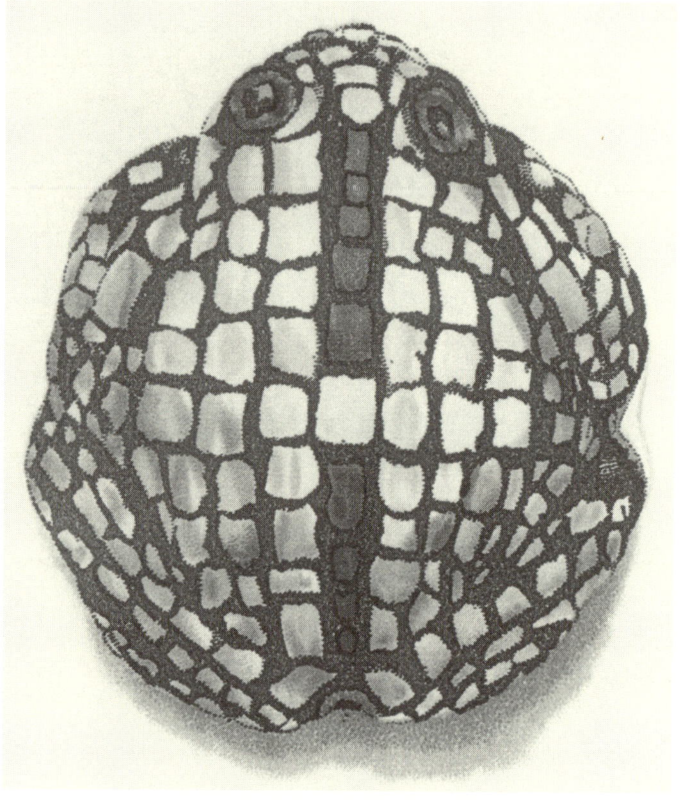

5.2. Painting, possibly by Margaret Magill, of the turquoise mosaic toad from Los Muertos that Cushing was accused by Hodge of faking (from Kunz 1890).

also as an assistant ethnologist. The first flurry of rumors died down, but in 1896, after he led a new expedition for the Archaeological Association of the University of Pennsylvania and the Bureau of American Ethnology to southwest Florida, where he discovered a marvelous series of sites and a stunning collection of wooden and other artifacts (Cushing 1897), fresh accusations of fraud were made. A Smithsonian photographer named William Dinwiddie accused Cushing of faking a painting on a sun shell (fig. 5.3) and intimated suspicions about other finds as well.[23] Hodge promptly renewed his claims about the turquoise toad, Fewkes joined in,[24] and thus began the worst scandal the bureau or the university had yet seen.[25]

Again the suspicions were suppressed, but not without consequences. Decades later, when he was writing up Cushing's Salt River valley collections for

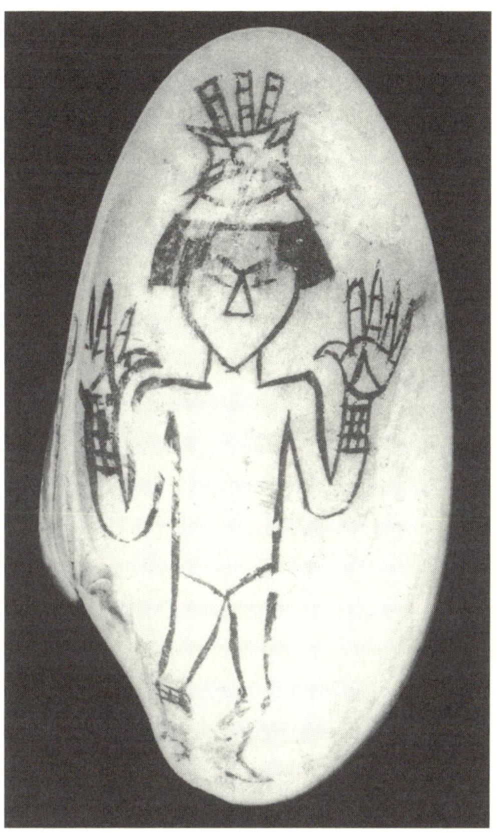

5.3. Sun shell from the Key Marco site with a painting that Cushing was accused by Dinwiddie of faking. Courtesy of the University Museum Archives, University of Pennsylvania, Philadelphia.

his dissertation at Harvard (Haury 1945, 1995), Emil Haury heard about the questionable nature of the turquoise toad and wrote to Hodge, who by that time was one of the grand old men of the "old archaeology." Hodge replied with a long account detailing his old charges (Hodge to Haury, October 5, 1931, SWM, MS 7, HAE 1.4). Based on his own observations of the specimen, Haury was strongly inclined to believe Hodge,[26] and William Sturtevant (personal communication) today continues to believe that the Florida shell is a fake. Such honest suspicions by major scholars seriously affect modern views of Cushing's contributions and, despite considerable discretion on the part of these scholars, have stained Cushing's anthropological reputation.

As an issue in the history of anthropology, Cushing's veracity is an apt focus for investigating what might have been if his vision of an ethnological anthropology had been realized. Beyond the question of whether he faked two (or more) artifacts are other questions about the veracity of his inferences and data about the American past and their resonance in the context of current scientific debates. To give only one empirical example—which I cannot resist: was it a ballcourt that he unknowingly recorded on Demerest's Key in 1895 (fig. 5.4) (SWM, MS 6: PHE 2.1)?

How are we today, more than one hundred years later, to resolve these issues? I suggest that we weigh the pros and cons of the extant evidence, as if in a court of law, and reach a judgment about these accusations using the standard of "beyond a reasonable doubt"—or any other we choose to invoke. We may thus achieve a consensus, but one that is subject to review as new ways to test the relevant alternative hypotheses are found. Such an investigation also provides us with a useful vehicle for avoiding "presentism" in seeking to understand how the profession of archaeology has come to be as it is.

Friends in High Places

At the Columbian Exposition in 1893, Cushing rebounded to the zenith of his powers, impressing two important participants in the emerging profession of anthropology, Zelia Nutall and Alice Fletcher, that he was "a sort of Sun in Anthropolog[y] illuminating the ways of all of them".[27] At the fair he began a lifelong collaboration on games with Stewart Culin, the director of the Archaeological Association of the University of Pennsylvania (Culin 1992), and he also met and favorably impressed Daniel Brinton, widely regarded at the time as the father of American anthropology (Darnell 1988, this volume).[28] Through Culin he drew closer to Brinton and identified with his advocacy of the autonomy of the American Indian Race (Wilcox 1999).[29] He also met Sara Yorke Stevenson, whom he met again that November in Washington, where she told him that she had spoken about him to "all Phila[delphia] my Friends—since Expos'n."[30]

After initiating a long-continued correspondence with Culin,[31] giving a well-received series of lectures at the Drexel Institute, in Philadelphia, and helping Culin to catalog and install the C. D. Hazzard cliff dweller collection, which they had first seen at the Chicago Fair,[32] Cushing came somehow to the attention of William Pepper,[33] the former provost of the University of Pennsylvania and the guiding light, with Sara Stevenson, of its museum. It was through these Philadelphia connections that it came to pass that Cushing

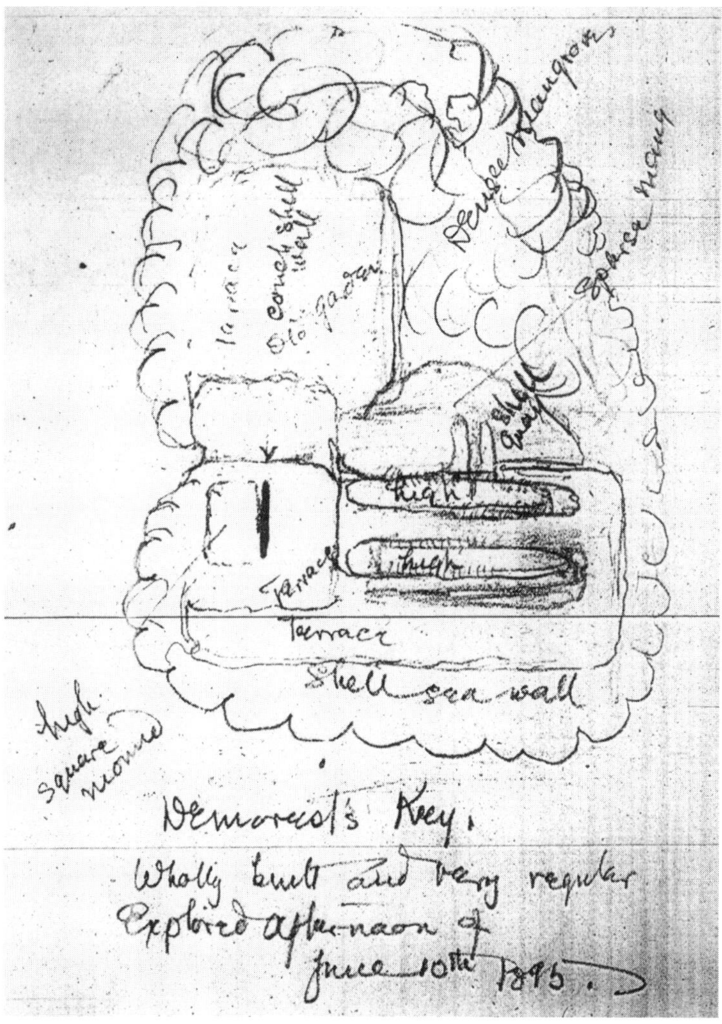

5.4. Possible ballcourt at the site of Demerest's Key recorded by Cushing during his 1895 reconnaissance. Courtesy of the Southwest Museum (Ms. 6:PHE 2.1), Los Angeles.

was in the right place at the right time to volunteer to investigate remarkable finds from southwest Florida reported to the University Museum. He soon talked Pepper into sponsoring a full-fledged expedition, and when the need to preserve the exceptional wooden artifacts arose, Pepper enlisted the support of Phoebe Apperson Hearst (Cushing 1897).[34] Upon his return from

Florida, on October 16, 1896, Cushing was inducted as a member of the American Philosophical Society.

After the scandal broke on October 5, 1896, Cushing presented his material for inspection at a meeting of the American Philosophical Society on November 6, and in the following year his preliminary report was published by the society in their *Proceedings* (Cushing 1897). John Wesley Powell had gone to Florida and seen the excavations firsthand (Cushing 1897) and would later tell Sara Stevenson that Cushing's analysis of the collections was a "work of genius."[35] Daniel Brinton and Frederic Ward Putnam both warmly discussed Cushing's report to the society and recognized the authenticity of the collections (in Cushing 1897).

Frustrated in his mission to expose fraud, William Dinwiddie, who had already tried going to the newspapers (see SWM, MS 6, No. 262), was fired, upon the strong recommendation of WJ McGee,[36] on November 6 by Secretary Langley for inattention to duty and insubordination.[37] Dinwiddie then wrote a long letter to Augustus Hemenway, laying out the charges being made against Cushing—and then asking for money.[38] With Mrs. Hearst's support, Cushing held back the newspapers with a threat of suit by her lawyers.[39] The accusers would not keep silent, however, and on February 6, 1897, Dinwiddie had the presumption to go to Hearst's home, but, Cushing reports in his diary, she "handled him fearfully" (Cushing diaries, February 6, 1897, NAA). Because of the virulence and persistence of the attacks against him, which included a sworn statement by Dinwiddie signed on February 15—and sent to Hearst[40]—Cushing on March 12, 1897, asked the secretary of the Smithsonian to initiate an investigation. But when Samuel Langley asked Powell about this, Powell set forth a strong defense of Cushing and stated that no investigation was necessary (cited in Gilliland 1975:181–82). Langley concurred, and Cushing so notified William Pepper.[41]

Thinking that that was the end of the matter, Powell quickly sought to move on, writing on March 24 to request Putnam's support[42] in arranging to have the Hemenway collections written up under his supervision.[43] Putnam, on April 6, 1897, did write to Augustus Hemenway, telling him that he had asked Cushing to give him a written statement about the turquoise toad,[44] which Cushing claimed he had restored. Putnam went on to say that "From all I could gather I do not think that Mr. Cushing intended to deceive in relation to this toad; but he was *injudicious in restoring the various parts.* This same method of restoration was followed with other specimens, and simply illustrates lack of archaeological training rather than any attempt at deception" [emphasis added].[45] Nothing came of this, however, Augustus Hemenway apparently preferring not to trust Cushing any further.

Was he right to be skeptical? Was Powell's failure to conduct an investigation a cover-up,[46] aimed at protecting one of his own against what he regarded as baseless jealousy, thus also protecting the bureau? The scandal perhaps had larger consequences, tearing apart the familial bonds within the bureau and weakening its support from the secretary of the Smithsonian. Langley asked not only Powell but also apparently a select "Committee for Promoting the Usefulness of the National Museum and Other Bureaus, Smithsonian Institution" to investigate. Dinwiddie's sworn statement was directed to that group, though what action it took I have not yet learned. Its head was Gardner Hubbard, the father-in-law of Alexander Graham Bell, the financial genius behind the Bell Company, and a founder of the National Geographic Society—in short, one of the most powerful cultural brokers in Washington, D.C. (Green 1962:100).[47]

Once Powell died in 1902, Langley acted, replacing Powell's handpicked successor, WJ McGee, with William H. Holmes and reducing the power of the position from director to chief (Hinsley 1979, 1981). Was this due, in part, to the Cushing scandal and to variant perceptions of its significance? Langley did move Frederick Webb Hodge close to him as acting chief clerk, where he would have had access to Langley's ear (Curtis Hinsley, personal communication).[48]

The Turquoise Encrusted Shell Toad

Hodge's charges regarding the turquoise toad are that Cushing saw a real specimen in the collections of a Phoenix man, determined that the Hemenway expedition must have one, and proceeded to make one "from scratch." He then palmed it off on the scientist Edward Sylvester Morse, a friend of Mrs. Hemenway, as a real specimen.[49] This story necessarily raises questions about Hodge's veracity as well as Cushing's, particularly as we have seen that he had motive. Was it a masterful lie, true in part and deceitful in crucial details?

There was a local collector, Lincoln Fowler, whose collections Cushing did probably see before the "discovery" of the Los Muertos toad.[50] Cushing's statement to Putnam[51] that the specimen he "restored" was "the first specimen of its kind recognizable as representing the desert frog (or toad) that had then been found" is difficult to reconcile with his previously having seen the Fowler specimen, which Hodge says was a frog.[52] However, in his diary, which he was directed to keep faithfully as a record of the expedition, on August 7 and again on August 8, 1887, Hodge comments that Cushing is "restoring" the turquoise toad.[53] There is no hint of fraud here. The area of the Los Muertos site being investigated at that time was Ruin XIV, and in both respects, this confirms Cushing's own detailed explanation to Putnam of how he found a real shell toad in a [Gila] polychrome jar that also contained what he thought was asbestos, in Ruin 4, later numbered Ruin XIV. He told Putnum that he made a drawing, and at the Brooklyn Museum there is a drawing of a partially intact turquoise toad (fig. 5.5),[54] which well corresponds with the "restored" specimen now in the Peabody Museum. Just when this drawing was made, or who made it, however, is unknown.[55]

In southwestern archaeology today, both turquoise-mosaic toads and raptorial birds are rare but recurrent finds in burial or other ritual contexts in Arizona and Chihuahua. A database compilation of these artifacts (Billideau 1986) has been supplemented to produce a data table, which includes fifty-nine cases (see Appendix). The bird forms first appear in the Flagstaff area during the Elden phase, A.D. 1150–1250, most famously in the so-called magician burial at Ridge Ruin (McGregor 1943). Toads also occur then, and both forms eventually are distributed from the Verde valley to the Phoenix and Tonto basins, Grasshopper Ruin, southeastern Arizona, and on to Casas Grandes in the fourteenth century. Most occur in burials. At Tuzigoot an important male burial had two toads as ear ornaments and a larger toad emblem at the throat.[56]

A rich burial at the late-thirteenth-century Pollack site had a bird emblem in its mouth. At Casa Grande ruins, two birds and a toad, together with much other turquoise and shell, were found in a cache in Compound A near the Big House, suggesting that these emblems of a nocturnal, earth-bound water symbol and a raptorial sky creature were the two aspects of a single dualistic whole. Quite possibly they were symbols of political office, which their occasional deposit with infants further indicates (see Wood and Wilcox 2000). These symbols were employed by several neighboring polities in central Arizona in the late Classical period in the Hohokam sequence (Wilcox et al. 2001).

Cushing's claim of finding a turquoise mosaic toad in a Gila Polychrome jar, which may also have held a ritual material (asbestos?), while unique, is entirely compatible with other known contexts for these special artifacts.[57] Emil Haury was bothered by scratches and other features of the specimen, particularly the bead indicating an anus.[58] Cushing's explanation of how he

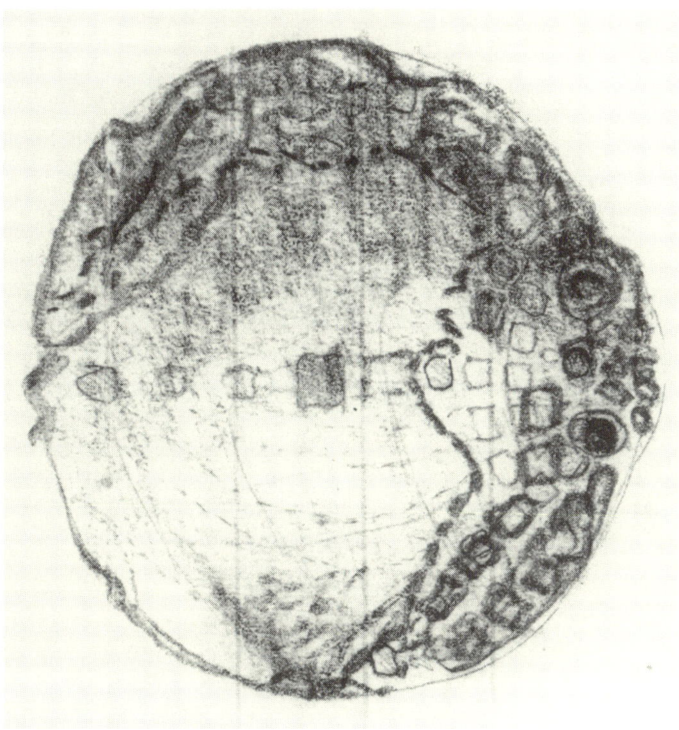

5.5. Drawing of a partially intact turquoise mosaic toad. Courtesy of the Brooklyn Museum of Art Archives, Culin Archival Collections, Cushing sketches, Hemenway Expedition (6.3.025 #1039), Brooklyn.

"restored" the specimen, however, adequately explains these observations. Cushing says the anus bead was original and was quite delighted when Jesse Walter Fewkes (1898) at Chaves Pass found another toad specimen that apparently had an anus indicated.

The stripe down the back is also unique. Most of these artifacts have a rectangle of red argillite on the back, which today is sometimes interpreted as representing the pararoid gland sac of the Colorado River toad, which contains a hallucinogenic substance (Joha 1996). Indeed, Davis and Weil (1992) have shown that *Bufo alvarius* (the Colorado River toad) does produce the chemical 5–MeO-DMT, which, when *smoked,* produces remarkable hallucinogenic effects (it would be interesting to see if any of the cane "cigarettes" in museum collections contain remnants of toad venom). While *B. alvarius* has a stripe, a more pronounced one appears on the back of *B. wood-*

housei (Behler and King 1979:225, 398–99), and unadorned shell-toad artifacts often exhibit stripes (Jernigan 1978:56, 68). A specimen from Elden Pueblo with the remnants of the lac glue, but lacking the turquoise tesserae, also exhibits a stripe. Comparing the "reconstructed" toad with the Brooklyn Museum drawing also suggests that Cushing may have erred in inferring that a stripe should be present. On balance, Putnam's judgment about this specimen seems sound.

One final piece of evidence comes from Sylvester Baxter, the secretary-treasurer of the Hemenway board, in a letter to Cushing dated May 2, 1897.[59] Baxter only arrived at Camp Hemenway on January 11, 1888, after the toad was "restored," but he *was* present when other artifacts Hodge questioned were discovered:

> I lately heard, only casually, that the *old charge* that you had manufactured the "jewelled toads" had been revived and was making a scandal for you, and that Hodge declared that you did. If I can be of any service to you in the matter I shall be glad to. The testimony of Drs. [Herman] ten Kate and [Jacob] Wortman ought to be of some value in the matter. I was with you when one of the [turquoise-inlaid] rings was found at the side camp where the so-called "llamas" [figurines] were found [at Los Guanacos]. [Edward Page] Gaston was there at the time, and I believe he is now on the Rocky Mountain News in Denver [emphasis added].

As Putnam (1897) and Hodge note,[60] Cushing "restored" some of these specimens too, which may partly account for the blackness of the lac in them all.

The Painting in a Florida Sun Shell

William Dinwiddie asserted that he became suspicious when Cushing pointed to a barnacle attached to a sun shell with an anthropomorphic painting, claiming that the superposition of the barnacle on the painting proved its antiquity. Looking at it under a glass, Dinwiddie observed that the reverse was true; the paint lapped up onto the barnacle. He then discussed his finding with others, including Hodge and Fewkes, and they told him of their conviction that Hemenway artifacts were fakes. Now fully aroused, but without authorization, Dinwiddie proceeded to have chemical tests performed, from which he concluded that the "paint" was India ink. He then took his charges of fraud to his boss, John Wesley Powell.

Marion Gilliland (1975, 1989) published two excellent books about Cushing's Florida expedition and, after a detailed examination of written testimony, concluded that the specimen is genuine. She shows, for example, that Carl Bergmann, the preparator of the Smithsonian, was present when the shell was found by another man, George Gause, the chief excavator, and described how it was opened up and the painting found, as well as how it became somewhat smeared when the mud and rootlets were removed (Gilliland 1975:179–84; 1989:96–14). This account fully substantiates Cushing's (1897) position (see also Gilliland 1975:183).

But what about the famous barnacle, the composition of the "paint," and the iconography of the figure? The barnacle has today disappeared, and clearly no line lies under where it once clung (Purdy 1996:Plate 54). How, though, would a barnacle grow *inside* a *closed* shell? Is it not likely that Cushing was wrong about the antiquity of the barnacle but not the painting? If the shell selected to be painted already had the barnacle on it, as now seems evident, then the overlap of the paint may have been applied aboriginally and is no proof of fraud.

Cushing (1897:387) thought the "paint" was probably rubber. Supposing that Dinwiddie's unauthorized chemical tests did not compromise the integrity of the paint, and if a small scraping would be allowed for destructive analyses, perhaps the issue of rubber or India ink could be scientifically settled, and an AMS date could be obtained that addresses its antiquity.

To learn more about the iconography, I showed the image to Edwin Wade, an expert in North American Indian art. He told me that the specimen is unique. Asked if it looked at all southwestern, he said that *in no feature* did it fit within southwestern art traditions. The hands with the "hitchhiker thumbs" reminded him of Adena-Hopewell styles, but from a local tradition. More extended comparisons would be desirable.

The temporal provenience of the all-but-unique Key Marco assemblage has long been a matter of great uncertainty. Cushing (1897) argued that this art spread from a southern, coastal source to the northern, Mississippian area, while later scholars view it as related to the late Mississippian Southern Cult (Goggin and Sturtevant 1964; Phillips 1973).[61] Recently, however, a date on the wooden board with what Cushing (1897:384, 426–27) thought was the polychrome image of a mythological bird-deity clutching a raccoon earth god produced an uncalibrated range of A.D. 600–720 (Purdy 1996:Plate 26).

Together with other linked traits to late Glades I assemblages near Lake Okeechobee in south-central Florida (McGoun 1993:106–7), current evidence leads Barbara Purdy (1996) to assign much of the Key Marco assemblage to the A.D. 600–800 period. Widmer (1989:179–80, cited in McGoun 1993:19) argues for a "South Florida Ceremonial Complex" related initially to the Hopewell ceremonial complex and "incorporating various Southeastern Ceremonial Complex motifs and traits into an indigenous religious system." So Cushing's intuitive sense of the position of the Key Marco assemblage may be amazingly on target.

Cushing suggested that the pointed face, without a mouth, signified a mask and that the narrow rectangular elements in the headdress were hairpins (not feathers). The wooden masks he found at Key Marco, he argued, represented animal concepts, but he made no suggestion about what animal was represented in the painting's "mask." Looking through Phillips and Brown's (1978) compilation of Southeastern Ceremonial Complex imagery, we find that the raccoon has a pointed face with no mouth showing and slanted eyebrows similar to the Key Marco shell painting (Phillips and Brown 1978: 136). The triangular nose is a simplification of the noses carved on other Key Marco masks (see Cushing 1897:Plate XXXIII; Gilliland 1975). There is now little reason to doubt the veracity of this specimen or the integrity of its discoverer.[62]

Discussion

William Dinwiddie may have been an honest whistle-blower confused by an error of logic and fooled by the slanderous allegations of people he trusted. A pawn of others, he paid the price for his insubordination with his dismissal. He went on to a career that may have better suited him: journalism (*New York Times* 1934). Quite the opposite happened in the careers of Hodge and Fewkes. Hodge became the editor of the *American Anthropologist* for fifteen years and remained on the editorial board long years after that. He became chief of the Bureau of American Ethnology, went on to a decade and more with the Heye Foundation, a bastion of the old archaeology, and ended up as the director of the Southwest Museum from 1931 to shortly before his death in 1956 (Judd, Harrington, and Lothrop 1957). Fewkes served as chief of the BAE as well and died in 1930. Both rose to fame by trampling on the reputation and discrediting the ideas of the "Sun" of a previous era, Frank

Hamilton Cushing. Were they liars? Let us say that they may have been unscrupulous in pursuit of their ambitions for fame and power.

More than that, they had different values than Cushing. Their conception of how archaeology should serve American society differed from his. Daring to be an eccentric (Evans 1997), Cushing used the pageantry of eccentricity to teach about the humanity of the "other," his friends the Zuñi (fig. 5.6). The Philadelphia realist painter, Thomas Eakins (Johns 1983:Plate 14) captured the contradictions of Cushing's persona in his famous 1895 painting of him in Zuñi costume. His tortured face shows the price he paid for his idealism. We could say that Cushing sought an emic understanding where Hodge and Fewkes knew only the possibility of etic descriptions (see Hinsley 1983a). Their conflict with Cushing may have been intensely personal, at least in Hodge's case, but it was also philosophical and value-laden, as were Cushing's conflicts with the two Stevensons, Matilda Coxe and Sara Yorke. The latter was intimately in tune with the civic culture of elite Philadelphia, and she helped build the University Museum to achieve cultural ends.[63] By 1899, she found Cushing to be a "hard case" and successfully turned Phoebe Hearst against him.[64] She later forced out Cushing's friend Culin, driving him from Philadelphia to Brooklyn, where he had a long and distinguished career.[65]

The story here, as yet unanalyzed and largely untold, is about the relationships of museum institutions to multiethnic and multileveled American society during the "confident years" (Brooks 1952) of the "genteel tradition" (Brooks 1915, May 1957, Wilson 1967), the era of the old archaeology (but see Cole 1985, Harris 1990, Conn 1998). The interesting story yet to be told (but see McVicker 1999, Snead 1999, 2001) is about the conflicts of values centered on these institutions and how university-based scholars with "transnational" (Bourne 1964) values achieved dominance in many of them in subsequent generations, the era of the first "new archaeology" (see Fowler and Wilcox 1999). There is much yet to learn about the history of anthropology and archaeology.

Acknowledgments. This work derives from a collaboration over many years with Curtis Hinsley for which I am most grateful. Jesse Green provided us with copies of the Cushing diaries, which I transcribed and have now deposited, with the assistance of Robert Leonard, with the originals in the National Anthropological Archives, Smithsonian Institution, Washington, D.C. I would particularly like to thank Michael J. Fox and Edwin L. Wade,

5.6. Cushing with three Zuñi friends at Manchester-by-the-Sea, Autumn 1886. Courtesy of the Peabody & Essex Museum, Salem.

Museum of Northern Arizona, for their support. Special thanks goes to Anthony Polvre and Dan Boone, Bilby Research Center, Northern Arizona University, and Tony Marinella, Museum of Northern Arizona. The following individuals and institutions over many years have been most generous in their cooperation with this research program: Mary Davis, Huntington Free Library, Bronx; Deirdre Lawrence and Deborah Wythe, Brooklyn Museum, Brooklyn; Kim Walters, Southwest Museum, Los Angeles; Steven LeBlanc, Peabody Museum, Harvard University, Cambridge; Allessandro Pezzati, University Museum Library, University of Pennsylvania, Philadelphia; Peabody-Essex Museum, Salem; Bancroft Library, Berkeley. David Gregory has provided many insights about the turquoise-mosaic toads and birds. I am especially grateful to the late Emil W. Haury for his encouragement and to William Sturtevant for his patience

Notes

1. One set of war gods that Cushing repainted is in a German museum, and another at the Pitt-Rivers Museum in England he allegedly made. The Zuñi regard the information as what is most important about the objects (T. J. Ferguson, personal communication, 2000).

2. Cushing told Mrs. Hemenway that "It is not the 'Science,'" it is "the *Poetry*, about my dear subject which makes me love it,—better than life, and far better than fame." He went on to say: "It is characteristic of Savage or Primeval thought that it reversed the Nature *we* conceive. It insisted on *Subjectivizing* every object, and objectivizing every Subject and on always personifying [im]personalities. This made a Myth of every thought and a Poem of every thing. My[thology], then, constituted the Religion and Philosophy of Primaeval man. Thus, it is only by the aid of the imagination, of the sympathies of poetic conception that we can wholly understand the [In]stitutions, the doings, even the Things of Primitive man!

"It is characteristic of the present school of thought of Ethnology that it, too reverses the Nature it studies. It insists on unsubjectivizing its objects, and unobjectivizing its subjects and on materializing *all* of it <even mythic and Poetic,> personalities. The result is this; as far as I can perceive, it succeeds only in proving itself to be in the Savagery period of its development; and its out-come is an irrationalism and darkness fully compatible to that of the Savagery it in so typically [in] a savage manner attempts to study" (Cushing to Mary Hemenway, June 1, 1888; Huntington Free Library (HFL), Cushing Letter Book (CLB) 4:341–49; see also Cushing 1895:310).

3. William Torrey Harris, for example, told Cushing that, "you cannot defend yourself from the almost universal charge that is made and will be made against your work, that is visionary, rather than scientific" (Harris to Cushing, November 9, 1891; SWM, MS 6 BAE 1.45; see also Kroeber 1930–35). Harris was U.S. Commissioner of Education, 1889–1906 (*Dictionary of American Biography* [DAB]).

4. In a review of this essay, Stephen Williams makes a case for the earlier successes of the Peabody Museum in stimulating public interest in American archaeology. There is merit in this argument (Cushing, for example, did look to Putnam's methods for guidance), but little of what Putnam achieved was widely institutionalized until the time of the Columbian Exposition, as the career of one of his "correspondents," C. C. Abbott (see Meltzer, this volume), well illustrates (see also Hinsley, this volume).

5. Nathan Reingold (1991:24–53) has stressed the importance of both certification and accomplishment in the emergence of professional organizations. Cushing lacked any college degree, but when he was made a member of the American Philosophical Society (see below) that was a powerful form of certification that may have helped him to overcome the charges made against him.

6. Mary Hemenway was no ordinary patroness. Two competing ideological movements in the nineteenth century guided American businessmen: Herbert Spencer's evolutionism and Hegelian philosophy (Cochran and Miller 1961). Hemenway supported the principal exponents of both: John Fiske for Spencer and William Torrey Harris for Hegel.

7. Cushing to Mary Hemenway, November 12, 1886, Southwest Museum (SWM), MS 6 Hemenway Archaeological Expedition (HAE) 1.28.

8. Mrs. Hemenway's goals can be known only indirectly, through the admonishments of William Torrey Harris to Cushing (May 9, 1889, SWM, MS 6 HAE 1.25). We might also note that the idea that Cushing begin the expedition with a reconnaissance came from John Wesley Powell (Cushing to Augustus Hemenway, letter, November 1886, SWM, MS 6 HAE 1.27), as did the idea that the focus for a report should be a daily itinerary (Cushing to W. T. Harris, November 22, 1886, SWM, MS 6 HAE 1.25). Cushing began the itinerary, but in May 1893 when Augustus Hemenway, Mary's son, refused further payment, he stopped (Wilcox 1999).

9. Sylvester Baxter to Cushing, November 18, 1886, SWM, MS 6, Bureau of Ethnology 1.6.

10. Missouri Historical Society, Harris Papers, Harris diary; Cushing to Mary Hemenway, July 18, 1888, HFL/CLB 4:496–99.

11. Cushing to Charles Garlick, July 22, 1889, SWM, MS 6 HAE 1.21.

12. Garlick to Emily Cushing, August 1889, HFL/CLB 2.

13. Cushing to Mary Hemenway, October 11, 1888, HFL/CLB 5:468–75; Hemenway to Cushing, telegram, June 15, 1889, SWM, MS 6 HAE 1.27.

14. Cushing to Hodge, March 31, 1889, SWM, MS 6 HAE 1.31.

15. Cushing to Pepper, November 1896, SWM, MS 6 Phoebe Hearst Expedition (PHE) 1.21.

16. Cushing to Don Carlos (Charles Garlick), February 17, 1889, HFL/CLB 6:85–89.

17. This letter is the only one missing in a series that leads up to it, but Cushing's draft of a long reply is extant, showing us what was in the missing letter (Cushing to I'ki na [Margaret Magill], March 30, 1889, HFL/CLB 6:141–56; Cushing to Hodge, March 31, 1889, SWM, MS 6 HAE 1.31).

18. August 16, 1887, Hodge diaries, SWM, MS 7, HAE 2.1; and Cushing to I'ki na (Margaret Magill), March 30, 1889, HFL/CLB 6:141–56.

19. On February 22, 1888, Hodge considered resigning because of the llama "affair" (Hodge diaries, SWM, MS 7, HAE 2.1).

20. Cushing diaries for 1890 and 1891, National Anthropological Archives (NAA), Smithsonian Institution, Washington, D.C.

21. Cushing diaries for 1891, NAA, Smithsonian Institution, Washington, D.C.; Ezra Coann to Talcott Williams, July 25, 1891, SWM, MS 6, Bureau of American Ethnology (BAE) 1.15; Cushing to Baxter, September 9, 1891, HFL/CLB 7:458–61; Cushing to W. T. Harris, September 10, 1891, HFL/CLB 7:467–69; Coann to Cushing, December 5, 1891, HFL/CLB 7:550–53.

22. Cushing to Don Carlos (Charles Garlick), March 30, 1892, HFL/CLB 7:649–59; see Kunz 1890.

23. Dinwiddie, in his accusations, repeated ideas he had about how science should be conducted. It is interesting to find in Cushing's diary for October 2, 1893, the following passage: "Had long talk at lunch time with Mr. Dinwoodie [sic] on Science[,] philosophy and religion—which did him good I think & did me good" (NAA, Smithsonian Institution, Washington, D.C.).

The fact that a layman like Dinwiddie felt perfectly comfortable asserting his own views about what science is shows the state of professionalization of archaeology in the 1890s.

24. Fewkes to Augustus Hemenway, November 30, 1896; File 4372(3), BAE correspondence (cited in Gilliland 1989:102).

25. "There has been a conspiracy here—which has ended by creating the worst scientific scandal we have ever known. My integrity has been distinctly questioned by men who professed friendship until the last. . . . The Bureau itself—its chiefs Major Powell and Professor McGee—are jeopardized" (Cushing to Pepper, November 1896, SWM, MS 6, PHE 1.21).

26. Emil W. Haury Papers, Arizona State Museum Archives, University of Arizona, Tucson, in the file on "Frog, Turquoise Encrusted."

27. Cushing 1893; Wilcox 1999; Cushing's diaries, August 30, 1893, NAA, Smithsonian Institution, Washington, D.C.

28. Brinton was memorialized as the "Founder of American Anthropology" after he died in 1899 (*Journal of American Folk-Lore,* 1900, 13:152).

29. Much to Powell's satisfaction, this involved a denial of Edward Tylor's claims for diffusion of backgammon to the Americas (see Cushing 1895; Cushing to Culin, November 27, 1894, Brooklyn Museum of Art Library, Culin Archival Collection [BMA/CAC], Cushing correspondence, 6.1.002).

30. Cushing diaries, November 29, 1893 (NAA). Sara Yorke Stevenson (1847–1921) was the vice-president of the jury of awards for ethnology at the fair. A founder of the Archaeological Association of the University of Pennsylvania, which soon became the University Museum, she was its secretary for ten years and then president of its board of managers. A woman with wide-ranging intellectual interests, she was elected to membership in the American Philosophical Society and was the first woman given an honorary degree by the University of Pennsylvania (DAB).

31. I have transcribed this correspondence, in the Brooklyn Museum of Art, Culin Archival Collection, and the University Museum, Philadelphia, onto disk and have placed the transcriptions with the originals. A book on the Cushing/Culin friendship is planned.

32. On Cushing's Drexel Institute lectures, see Cushing-to-Culin correspondence, BMA/CAC, CC 6.1.002–003. On Cushing's assistance with the Hazzard collection, see Administrative Records, American Section Curatorial, Box 25, Folder 13, University Museum Archives, University of Pennsylvania, Philadelphia.

33. Just how that happened remains unclear. Cushing (1895) states, however, that he was in the hospital at the University of Pennsylvania under Dr. Pepper's care. While he had managed to expel the tapeworm he had in 1890 (see Cushing to Washington Matthews, February 20, 1890, HFL/CLB, 7:10–14), Cushing still suffered from diverticulitis owing to a deformed stomach and required special diets and sometimes the care of physicians (Cushing diaries, NAA, Smithsonian Institution, Washington, D.C.).

34. Cushing may have first met Hearst when he visited San Francisco in 1887; shortly thereafter Sylvester Baxter (HFL/CLB 3:insert, pp. 177–78) reported to Mrs. Hemenway, January 20, 1888, on a scheme for William Randolph Hearst to loan his yacht to Cushing for an archaeological survey along the West Coast as far south as Peru. When Cushing was in Washington lobbying to have the Casa Grande set aside, Mrs. Hearst opened her home for this effort (Cushing to Pepper, April 7, 1896, University of Pennsylvania, Museum Archives [UPM]).

35. Powell to Stevenson, February 3, 1899, UPM, Expedition Records, North America, U.S., Florida Explorations, Box 2, Folder 9, Cushing Post-Excavation Correspondence 1899–1900.

36. McGee to Langley, 27 October 1896, Barnett Papers, NAA, Box 3.

37. Langley to Dinwiddie, November 6, 1896, Smithsonian Institution, Series 34, Secretary, 1891–1907; Outgoing Correspondence: Ethnology. Box 23: Folder 4.3, January 2–December 26, 1896.

38. Dinwiddie to Hemenway, November 18, 1896, Peabody-Essex Museum, Hemenway Collection (PEHC), Box 9.

39. Cushing to George W. Rowzer, November 9, 1896. SWM, MS 6, PHE 1.29, cited in Gilliland 1989:101.

40. Dinwiddie to Hearst, February 15, 1897, Bancroft Library, Phoebe Hearst Collection 72/2042, Box 16.

41. Cushing to Pepper, March 22, 1897, SWM, MS 6, PHE 1.22.

42. A letter from Frances Mead to Alice Putnam, also on March 24, 1897, states that, "It is in connection with this question [about the turquoise toad] that the Hemenway estate representatives have asked your father to go to Washington . . . *They perhaps begin to feel that they have been unjust and wish to find out the truth if possible.* . . . He is simply in charge now of the whole Hemenway collection because it has been arranged in this Museum [i.e., the Peabody Museum, Harvard]. Augustus Hemenway is one of our Visiting Committee, and seems much interested in the Museum" (emphasis added; Ralph W. Dexter Papers, Kent State University Archives, Kent State,

Ohio; notes taken by Curtis Hinsley, August 24, 1994). Indeed, in later years Hemenway was a patron for Peabody expeditions (Peabody Museum archives), also supporting the work of the Boston Fine Arts Museum in Egypt (see Whitehill 1970).

43. Powell to Putnam, March 24, 1897, PEHC Box 9.

44. Putnam to Hemenway, April 6, 1897, Peabody Museum, Harvard University.

45. Putnam went on to mention that similar "restorations" had been made of remains from classical lands. Indeed, the first professional director of the Metropolitan Museum of Art in New York, Emmuele di Cesnola, was sued for slander in connection with a charge by a newspaper reporter in 1880 that he faked a sculpture of Aphrodite (by adding a mirror in its hand). Defended by the most powerful lawyer in New York and a longtime trustee of the Museum, Joseph Choate, Cesnola beat the rap (Tompkins 1970).

46. On August 3, 1973, near the height of the Nixon scandals, Emil Haury wrote to Stephen Williams at the Peabody Museum asking for a copy of the catalog card (for specimen 585a) pertaining to the frog, adding that it was "an obvious hand written insertion in the system—an 1888 cover-up!" Haury's theory of the "faking" was that Cushing made a model that was then taken as the real thing, and, rather than admitting that, he proceeded to elaborate a hoax (Emil W. Haury Papers, "Frog, Turquoise Encrusted," Arizona State Museum Archives, University of Arizona, Tucson).

47. Gardner Greene Hubbard (1822–97) was also a founder of *Science,* became a regent of the Smithsonian in 1895, and from 1895 until his death in December 1897 was the president of the joint commission of the scientific societies of Washington (DAB).

48. This is a matter for further research. The annual report of the Smithsonian Institution for 1896 already shows Hodge as acting chief clerk, although he may already have held this position in 1895 (Curtis Hinsley, personal communication). But what influence, if any, did he have over Langley's judgment?

49. Hodge to Haury, October 5, 1931, SWM, MS 7, HAE 1.4.

50. Lincoln Fowler (1859–1924) became one of "the leading spirits in the development of the Salt River Valley, becoming the prime mover in building the 'Appropriators' Canal' to save crops under the Grand Canal after the dam of the Arizona Canal was washed out in 1891" (*Arizona Republican,* September 2, 1924, p. 14, col. 5; September 3, 1924, p. 12, col. 2). In 1887 he was a

Phoenix City councilman from the fourth ward; he drew up the plans for the first Phoenix City Hall (*Phoenix Herald,* May 5, 1887; September 19, 1887, p. 3, col. 3). See Fowler to Cushing, May 19, 1887 and Fowler to Cushing, October 26, 1889 (SWM, MS 6, HAE 1.20).

51. Cushing to Putnam, June 1897, SWM, MS 6, BAE 1.95.

52. Hodge to Haury, October 5, 1931, SWM, MS 7, HAE 1.4. Possibly the Fowler specimen was recognized as a "fetish" (as Fowler says to Cushing) but was not immediately recognizable as a toad symbol. Some of the "toad" specimens are only recognizable as such by analogy with other of the like specimens.

53. On August 7, Hodge says: "Mr. Cushing has busied himself to-day by restoring a shell which was originally inlaid with mosaic." On August 8: "Mr. Cushing held a camp-meeting in my tent this evening for the purpose of exhibiting a very handsome shell frog-fetish which he has been engaged in restoring during the last few days" (SWM, MS 7, HAE 2.1).

54. The drawing of a turquoise-mosaic toad is in the Brooklyn Museum of Art, Culin Archival Collection, Cushing sketches, Hemenway Expedition, 6.3.025 #1039.

55. For all we know, it might be a drawing of the specimen Cushing saw in the Lincoln Fowler collection. This drawing is probably at the Brooklyn Museum of Art because it was part of Cushing's library and manuscripts, which Mrs. Cushing gave to Stewart Culin shortly after Cushing's death in 1900, and which he took with him to Brooklyn in 1903. Culin in 1921 gave Fred Hodge, who was then at the Heye Foundation, three crates of these manuscript materials, and thus the Huntington Free Library and later the Southwest Museum came to have their Cushing collections (see Hinsley and Wilcox 1995).

56. See photograph in Simmons, n.d., Archives A-402, p. 28, Arizona State Museum, University of Arizona, Tucson.

57. At my request, Steven LeBlanc, director of collections at the Peabody Museum, examined the toad specimen and the Gila polychrome jar. There is a hard, white substance in the base of the jar, but it is not clearly asbestos. Further testing is needed. The toad fluoresced under ultraviolet light differently than all but one ring mosaic specimen. To see if there is original lac under the "restored" material would necessitate taking the specimen apart. To evaluate fully the scientific integrity of this specimen would require just such an intrusive and destructive examination. Perhaps someday that can be done.

58. Other unusual features that Haury noted included the lack of edge-trimming on the turquoise pieces, their ragged edges, their awkward fit together, and the shell beads representing eyes and anus (EWH Papers).

59. University of Pennsylvania, Museum Archives (UPM), Expedition Records, North America, U.S., Florida Explorations, Box 2, Folder 7, Cushing Post-Excavation Correspondence 1897.

60. Hodge to Haury, October 5, 1931, SWM, MS 7, HAE 1.4.

61. Cushing (1897) also argued that his Florida finds represented trans-Caribbean influences that affected the rest of the eastern woodlands. Recognizing that archaeology is fundamentally the study of relationships and viewing his ideas as a pioneering intellectual effort at synthesis and an attempt to relate his material to the highest level of debate about American Indian culture, the realm of Daniel Brinton and Frederic Ward Putnam, it would be interesting to compare his hypotheses to the recent ideas of William Sears (1977, 1982) or Thomas Riley (1987) to see just how prescient Cushing was, albeit still controversial! Happily, the University Press of Florida has reissued Cushing's Key Marco report with an introduction by Rudolph Widmer.

62. By the time of his death, Cushing had nearly finished a "voluminous report" on his Florida work, according to Powell (in *American Anthropologist* 1900:366; see also what is probably an obituary by Culin [Anonymous 1900]). At the Southwest Museum there are numerous topical fragments of such a manuscript, and one long typed manuscript, corrected in Cushing's hand, entitled "Remains of Ancient Key Dwellers on the Gulf Coast of Florida" (SWM, MS 6; PHE 4.5, 31–262). In the Culin Archival Collection at the Brooklyn Museum of Art Library are numerous drawings and figures that probably pertain to it. How fascinating it would be to see what ideas Cushing came to in his extensive comparative studies!

63. Wister 1922, Myerson and Winegard 1978, Danien and King, this volume.

64. Stevenson to Hearst, June 9, 1899; Stevenson to Sparhawk, July 3, 1899, UPM; Cushing's position was explained to Culin on August 10, 1899, BMA/CAC, CC 6.1.005.

65. For Culin's Brooklyn career, see Lyman 1982; Bronner 1989; and BMA/CAC.

6

Clarence Bloomfield Moore

A Philadelphia Archaeologist in the Southeastern United States

Lawrence E. Aten and Jerald T. Milanich

Clarence Moore, who probably excavated and reported more archaeological sites than any individual who ever lived, is practically unknown outside the circle of archaeologists working in the southeastern United States. Moore was a lifelong resident of Philadelphia who over the period 1891 to 1918 excavated more than 850 sites, the vast majority of which were mounds. He was never a member of the clique of anthropologists who dominated the field at the turn of the century from their university and museum settings in Cambridge, New York, Philadelphia, and Washington. Instead, he was one of the last of the nineteenth century's free-wheeling independent researchers. By undertaking yearly field expeditions and faithfully reporting his findings, he defined most of what for many years was known about the material culture of southeastern mortuary ceremonialism. And he constrained, by the extent of his digging, what can ever come to be known about this subject.

Moore was a prolific writer. The core of his archaeological reportage, largely published in the Academy of Natural Sciences of Philadelphia's *Journal,* consists of thirty-five quarto-sized reports, some monograph-length, richly illustrated with artifact drawings, watercolor paintings, and photographs (a list of Moore's *Journal* reports is reproduced as the frontispiece in the last issue in which he published; see Moore 1918). In addition, articles about his early 1890s work in Florida were published in the *American Naturalist* (Moore 1892a, 1892b, 1892d, 1893a, 1893b, 1893c, 1894b, 1894c) and the *American Antiquarian and Oriental Journal* (1892c). In 1896 he privately published three additional papers on his work in northeast Florida, all of which were reprinted in 1922 in the Museum of the American Indian, Heye Foundation's *Indian Notes and Monograph* series (Moore 1922). In 1902 Moore published a very fine, but not well-known, summary of his previous decade's work in the *Proceedings of the 13th International Congress of Americanists* held in New York (Moore 1902).

Subsequently six other Moore articles on southeastern archaeology were published in the *American Anthropologist* (Moore 1903a, 1903b, 1904, 1905a, 1919, 1921a). The first of these, which included comments of discussants, definitively summarized his landmark studies on the origin of sheet copper artifacts found in southeastern mounds, a subject of some controversy at the time. A version of his 1921 article on Florida shell tools was published in the inaugural issue of *Boletín de la Academia Nacional de Historia* (in Quito) the same year (Moore 1921b).

Not all his archaeological publications focused on the southeastern United States. "The Boro Budur Temple of Java" appeared in 1899 in the proceedings of the Numismatic and Antiquarian Society of Philadelphia for the years 1892–98, and "Likenesses in Paleolithic Relics" was published in the *American Antiquarian and Oriental Journal* in 1894. His 1904 *American Anthropologist* article on urn burial also sought to review this burial practice as it occurred throughout the United States. Later the *American Anthropologist* contained two short articles by Moore on the Red-paint People of Maine (Moore 1914, 1915b).

In addition to his prodigious publication record, Moore's field notebooks, curated at the Huntington Free Library in New York City (Davis 1987), and his excavated collections are still in existence. The bulk of the latter now are curated at the Smithsonian Institution's National Museum of the American Indian, although additional materials remain in the collections of museums in America, Europe, and the Far East. His researches today provide archaeologists with baseline data on Woodland and Mississippian period cultures, especially those of Florida (e.g., Goggin 1952:33–34; Rouse 1951:63–64; Willey 1949:21–26), but most other southeastern states as well (e.g., Lewis 1996:xi, 8–9; Lewis and Kneberg 1970; Stoltman 1973:128–32; Walthall 1980; Williams 1977; Williams and Brain 1983:11–12; Moore also excavated at two important archaic sites, Poverty Point in Louisiana and Indian Knoll in Kentucky). The reprinting of Moore's *Journal* reports by the University of Alabama Press further emphasizes their continuing significance a century later.

Moore's Family and Early Life

Born January 14, 1852, in Philadelphia, Clarence Bloomfield Moore was the son of Bloomfield Haines Moore and Clara Sophia Jessup. The Moores were Quakers from New Jersey; the Jessups were Massachusetts Puritans; and both had become established in Philadelphia. Clarence's paternal grandfather was the well-known early-nineteenth-century scientist Augustus Edward Jessup, who had been a geologist on Stephen H. Long's 1818–19 expedition of discovery to Yellowstone Territory, Wyoming (Kelly 2000). In 1841 Jessup's daughter Clara married Bloomfield Moore (Moore 1940). Two years later Bloomfield and his father-in-law established Jessup and Moore, a paper manufacturing company that prospered over the remainder of the century (Scharf 1888:2:793–96).

Although they belonged to Philadelphia's new industrial rich and not the even more exclusive patrician social class, the Moores were socially prominent and highly respected (Lewis et al. 1987:7). They raised three children—Ella, Clarence, and Lillian—and were active in numerous civic activities. Both Bloomfield and Clara were avid collectors, the former of books and the latter of paintings and sculpture (Lewis et al. 1987:62; Scharf 1888:2:794).

Early in their marriage, while Bloomfield was building the family company, Clara began a successful writing career. From the 1840s she was writing and publishing novels, short stories, children's stories, and poetry. Her books on etiquette were the most popular of her writings; *Sensible Etiquette of the Best Society,* first appearing in 1878, went through at least twenty editions. Clara Moore also was a strong supporter of higher education for women (Garraty and Carnes 1999:741–42).

Young Clarence had a privileged upbringing of wealth, socially prominent friends, and private schools both in Philadelphia and Europe. His mother—ever the domineering personality—decided that Clarence should attend Harvard College, where his career was a quiet one (Moore 1940:6). Both Moore daughters married Swedish aristocrats and had, between them, seven children reaching adulthood. Several of these were very close to their Uncle Clarence; among these was Eric von Rosen, a prominent early-twentieth-century cultural anthropologist (Bohman 1949:6:334–35).

After graduating from Harvard in 1873 at age twenty-one, Clarence traveled in Europe, the eastern Mediterranean, and Egypt for nearly three years. This foreign travel in summers was punctuated with winters in Florida. In 1876 he traveled to Peru and made an arduous eastward journey across the

6.1. Clarence B. Moore posing informally during Harvard Class of 1873 picture-taking. Courtesy of the Harvard University Archives.

Andes and down the Amazon River. The next year, Moore began an around-the-world tour. In the summer of 1878, while Clarence was in Japan, his father died unexpectedly. Clarence returned promptly and, so far as is known, never traveled abroad again (*Paper Trade Journal,* September 21, 1878:304; *Philadelphia Inquirer,* July 6, 1878; Ware 1876:18, 1898:28).

Appointing his wife and son as co-executors, Bloomfield Moore bequeathed his entire estate to his three children but left his wife full discretion to determine when and what proportions each would receive. Meanwhile, his wife was directed to take the income from the estate to create her own "nest egg." Although the estate was appraised at $5.5 million, most of this value was encumbered in real property, equipment, and inventory of the unincorporated family company (Bloomfield H. Moore, Last Will and Testament Petition and Inventory. 1878, Will No. 489, Register of Wills, Philadelphia), assets certain to bring much less value if liquidated.

Competitors assumed the widow Moore, who already had a sizable fortune inherited from her father, would sell the assets of the company at bargain-basement prices. By this means she would obtain cash to take her "widow's third," to which she was entitled by Pennsylvania law, and then leave the children to divide the substantially reduced amount remaining from their father's estate (Duncan 1882).

Clara Moore had her own plans, however. She incorporated the company, with the estate holding the great majority of shares; she had Clarence installed as company president—probably as a reminder to everyone of who was in charge; and she reaped the income from the estate for the rest of her life. Although for a few years she was generous in distributing portions of the income (Duncan 1882), the assets of the Bloomfield Moore estate were not distributed to the heirs until nearly twenty-five years after his death.

From the early 1870s Clara Moore spent a great deal of time in Europe. She purchased a home in London, only a stone's throw from Buckingham Palace, and became a society figure. One of her close friends there was the poet Robert Browning (James 1971:2:573–74).

Clara Moore, who now had changed her name to "Clara J. Bloomfield-Moore" (Brown 1903:5:531) was the major patron of the Philadelphia inventor John E. W. Keely from about 1882 until he died in 1898. Keely claimed to have discovered an "etheric force" and invented numerous motors and devices driven by this force. Keely explained his theories in what could best be described as gibberish—made-up terms and concepts that no one understood.

Despite being labeled a charlatan by *Scientific American,* the charismatic Keely had many prominent supporters (Anonymous 1907:137–38). Of these, Clara Moore was the most steadfast, forthcoming, and articulate. She believed she understood Keely's theories, and in a desperate attempt to record his work in the event of his death she authored several articles as well as a 372-page book in 1893 (*Keely and His Discoveries*). This was an arduous task for her because her health had become quite fragile by this time. However, she continued her moral and financial support—the latter aggregating to about $100,000 by one estimate (*Press,* January 22, 1899:1).

This support fueled a growing estrangement between Clara and Clarence and was principally responsible for his refusal, after 1891, to adopt the University of Pennsylvania as the primary repository for his archaeological collection. Clara urgently wanted to find secure funding for Keely before she died. In addition to petitioning her many friends, she wanted, indeed she expected, Clarence to step up to the plate and lead this effort. Clarence, of course, thought Keely was a quack and would do nothing of the sort. As early as 1891, individuals associated with the University of Pennsylvania unwisely sided with Clara in this matter (C. B. Moore to S. Culin, letter, December 30, 1891, Administrative Records, American Section, Box 30, Folder 9, University of Pennsylvania Museum Archives). Clarence deeply resented this intrusion. Although various persons at the University were his friends, Clarence Moore always exhibited coolness toward the University Museum.

The mother-son dispute in time led to much unpleasantness. In 1895 Clara Moore petitioned the court overseeing administration of her husband's will to remove Clarence as co-executor citing his refusal to communicate with her about business matters. Clarence, in turn, petitioned the court to remove his mother on the grounds that she was insane (*Public Ledger,* December 2, 1895:16; January 20, 1897:13). The court appointed a master to take testimony from both sides, and in the middle of this Mrs. Moore left for London, never to return to Philadelphia.

John Keely died in Philadelphia in November 1898, and Clara Moore died in London a month and a half later. One of her many obituaries quotes her friends as saying she "died of a broken heart," her entire life being "centered in [Keely's] work, to the exclusion of all other interests" (*Public Ledger,* January 6, 1899:1). When news of his mother's passing arrived, Clarence was in Philadelphia for the holidays and was preparing to return to Mobile, Alabama, to resume his 1898–99 field season.

As the surviving co-executor of his father's estate, Moore quickly acted to liquidate the family's interest in Jessup and Moore Paper Company and to distribute to the heirs what remained of his father's estate. Once the sale was completed, in the late summer of 1899, Clarence Moore retired as president but continued on the board of directors (*Paper Trade Journal,* September 14, 1899:492).

He also was determined to get to the bottom of Keely and his "inventions" that had caused so much anguish with his mother over the prior twenty years. He leased the Philadelphia premises where Keely held his demonstrations and, under the supervision of Moore's friend Milo G. Miller, dismantled the building's interior. This disclosed that compressed air fed through hidden pipes ran Keely's machines, not harmonic vibrations releasing "etheric energy," just as Keely's detractors had predicted (*Press,* January 19, 1899:1). Even so, Keely today has an Internet following who rely heavily on Clara Moore's book for their understanding of "sympathetic vibratory physics."

A Developing Interest in Archaeology

As far as is known, Clarence Moore first collected archaeological artifacts in the 1870s. Although he was never trained as an archaeologist, it has been speculated that while a student at Harvard, Moore was in contact with the archaeologist Jeffries Wyman, Harvard Peabody Museum's first curator, and possibly was in the field with Wyman in Florida in 1873 (Murrowchick 1990:64). But neither Moore's lifelong correspondence nor Wyman's letters reporting on his Florida work in the 1870s to the chairman of the Peabody Museum's Trustees gives any hint of this. Moore mentioned Wyman from time to time in his correspondence and reports, but always in terms of knowledge from reading Wyman's published reports. If Moore had ever made personal acquaintance with Wyman it is practically inconceivable, given Moore's personality, that he would not have mentioned this.

Clarence Moore made chartered boat trips up Florida's St. Johns River six times from 1873 to 1882 and then did not return until 1890 (Moore 1987:Roll 6, Unnumbered notebook:19; 33). Using the Magnolia Hotel resort, near Green Cove Springs, as his base, the focus of these early recreational trips was shooting and fishing. While out hunting in 1873 his attention was drawn to a sand mound and, in the manner of budding enthusiasts everywhere, he uncovered a few artifacts by digging with a stick at the surface of the site (C. B. Moore to F. W. Putnam, letter, March 16, 1891, UAV677.38, Peabody Museum Correspondence, Box 11, Folder 1891 L-M, Harvard University Archives). In 1879 he spent the night at Bluffton on another hunting trip. Coming upon a sand mound there, he devoted a half hour with a sheaf knife to digging a small hole and found some artifacts (C. B. Moore to F. W. Putnam, letters, April 8, 1879, UAV677.38, Peabody Museum Correspondence, Box 2, Folder 1879 K-P; March 16, 1891, UAV677.38, Harvard University Archives). It would be a stretch to call this early activity "research," but Moore came from a home in which the urge to collect was rampant, and his inclinations were no different.

By the early 1880s three benchmarks were evident in Moore's nascent archaeological interests. First, he had established a pattern of not maintaining a private artifact collection but of donating all his artifacts to a public museum where they could be seen and studied. Second, his correspondence indicates that he was reading about archaeology. And third, this was the time his acquaintance with Frederic Ward Putnam began. Although this would not blossom into a friendship until the end of the decade, their later correspondence shows that they held each other in high esteem and great affection. It was Putnam who would become the nearest thing to a mentor to Clarence Moore's archaeological career.

From 1882 to around 1889, Moore apparently confined himself largely to Philadelphia. The details of these years are obscure, but much happened. In the early 1880s there seemingly was the first open conflict with his mother, and Clara Moore removed more or less permanently to London, leaving Clarence to act on her behalf in estate business matters.

It may have been at this time that Clarence realized he would have to accumulate his own fortune if he were to have one. With his father's estate undistributed, Clarence lived in his mother's great Furness-designed mansion at 510 South Broad Street—a "flower by the roadside" said architect Louis Sullivan (Thomas et al. 1991:63)—and devoted the decade to growing wealth (assisted for a time by gifts from his mother) rather than spending it. Although the maximum amount he attained is not known, one might estimate, based on his own estate's appraisal after his death (C. B. Moore, Last Will and Testament Petition and Inventory. 1936, Will No. 1122, Register of Wills, Philadelphia) and his focus on conservative investments in bonds, real estate, street railways, and so on, that the amount was between $1.5 and 2 million.

Meanwhile, Moore's cousin, J. Ridgeway Moore, who had charge of Jessup and Moore's New York office, interested Clarence in photography. He threw himself into this hobby, and by the time it was given up a dozen years later, he had become an often-published author and an award-winning amateur photographer (Finkel 1980:218; Panzer 1982:38). (Today the whereabouts of only six of his original prints are known.)

In 1887, during a game of tennis, Clarence took the ball squarely in his left eye, suffering a detached retina and near total loss of vision in that eye (C. J. Moore to C. Frothingham, letter, September 8, 1892, C. J. Moore Papers, 1890–95, E-13, #1867, Historical Society of Pennsylvania; Ware 1888:25). This painful accident profoundly affected the remainder of his life. After 1898, about the time his good eye began to trouble him as well, he turned away from photography. He seldom went out in evenings because he found the glare from lights too painful (C. B. Moore to E. J. Nolan, letter, October 10, 1897, Official Correspondence, Coll. 567, Academy of Natural Sciences, Philadelphia). He gradually limited his travel away from Philadelphia to his winter trips to the South. Over time he gave up writing in favor of dictation, and difficulties with his right eye came to interfere with his reading. This accident seems to mark the point in Moore's life when he began to withdraw from the outside world. Indeed, his physician at the time of the accident, Bowman Hendry—also a lifelong bachelor—moved into Clarence's house and lived there for the rest of his life (C. B. Moore to F. W. Putnam, letter, November 11, 1904, F. W. Putnam Papers, General Correspondence, Box 17, Folder 1901–10, Harvard University Archives).

At the end of the decade, Moore suffered some unidentified medical problem apparently necessitating extended medical care (C. J. Moore to C. Frothingham, letter, June 5, 1890, C. J. Moore Papers, 1890–95, E-13, #1867, Historical Society of Pennsylvania). This may have been the time when Moore and Milo Miller first became acquainted. Dr. Hendry had an established practice and could not be with Moore at all times. Dr. Miller, who had just graduated from medical school (Anonymous 1942), may have been hired to assist. In 1890 Hendry and Miller accompanied Moore to Florida ostensibly for recuperation. They traveled about the state engaging in photography, inspecting archaeological sites, and probably fishing a great deal as Moore was an avid fisherman. During this time Moore resumed corresponding with Frederic Ward Putnam about archaeological shell sites he had seen being quarried (C. B. Moore to F. W. Putnam, letter, December 12, 1890, UAV677.38, Peabody Museum Correspondence, Box 10, 1890 C-Z, Harvard University Archives).

In the following year, 1891, matters become clearer. Moore, accompanied at least by Dr. Miller, returned south to begin a two-and-a-half-month trip on the St. Johns River on the chartered steamers *Osceola* and *Alligator*. This trip was intended to be recreational, engaging in archaeological surface collecting, some very superficial digging, and occasional artifact purchases, along with a lot of rowing, fishing, and photography (Moore 1987:Roll 6, Unnumbered notebook:1–75).

However, early in the trip, Putnam wrote Moore to suggest he consider excavating an archaeological site (C. B. Moore to F. W. Putnam, letter, February 23, 1891, UAV677.38, Peabody Museum Correspondence, Box 11, Folder 1891 L-M, Harvard University Archives). "Agreeable to a wish,"

Moore focused on the large conical sand mound at Tick Island as the only "unopened" mound he could get permission to dig. He engaged a couple of men to help dig, kept written field notes of a sort, and raked through the displaced sand to recover bones and artifacts (Moore 1987:Roll 6, Unnumbered notebook:64–97). He reported his findings to Putnam, and a correspondence ensued in which there is a hint the latter may even have encouraged Moore to pick up where Jeffries Wyman left off. In any event, and inferring from correspondence (C. B. Moore to F. W. Putnam, letter, March 16, 1891, UAV677.38, Peabody Museum Correspondence, Box 11, Folder 1891 L-M, Harvard University Archives), Putnam apparently gave technical advice on how to proceed, and the rest, as they say, is history. Upon returning to Philadelphia that spring, Moore immediately wrote an account of this investigation and arranged, possibly through his friend, E. D. Cope, publisher of the *American Naturalist,* for its publication as "Burial Mound of Florida" (Moore 1892a).

This sudden transformation from tourist to practicing archaeologist was the epiphany of Moore's life; there is no indication it was premeditated. Clearly he had had an interest in things archaeological for all of his adulthood, but so do many people who never become archaeologists. And while his skill and experience underwent much further development, the 1891 work at Tick Island in Florida marks the beginning of Clarence Moore's thirty-year career as an archaeologist.

Getting to the Field

From 1890, Moore resumed regular winter trips to Florida, ostensibly for his health, and from 1892, he primarily occupied his time there with archaeology. Moore continued to suffer vision problems with his "good" eye, as well as gout, vertigo, severe digestive problems, and other illnesses that occasionally laid him up. These medical problems were no small part of his life, leading him to narrow further his functional environment. He once stated that he was quite timid about leaving his home and explained the obvious contradiction that when it came time to leave Philadelphia for the field, he went directly from his house to a Pullman car, and from there to his steamboat where he took up residence; "I am practically home all the time" (C. B. Moore to F. W. Putnam, letter, June 27, 1900, HUG1717.2.1, F. W. Putnam Papers, General Correspondence, Box 10, 1891–1900: G-M, Harvard University Archives). Though Moore seems to have been an adaptable, gregarious field worker, from the time of the 1887 eye accident he progressively built a protective cocoon around his life.

Much of Florida in those days was still wilderness, and transportation to some regions could be difficult at best. Moore solved the transportation and housing problems by using shallow-draft, paddle-wheel steamboats that not only facilitated access to sites but also provided sleeping quarters. Until mid-1895 Moore chartered three boats: the *Osceola,* the *Alligator,* and the *Metamora* (Moore 1987:Roll 1, Notebook 5:228; Mueller 1983; Pearson et al. 2000:83). A New York tourist mistakenly wrote that Moore owned the *Alligator* (Swift 1903:47), but Moore's journals and correspondence indicate that all the boats used in the early years were only chartered (Moore 1987:Roll 3, Notebook 19: following p. 253). The most famous of his steamers, and the only one he owned, was the *Gopher of Philadelphia,* which Moore had built at Jacksonville, Florida, in 1895, after deciding the *Alligator* no longer fit his needs. The *Gopher* served Moore as a water-based field station for the rest of his archaeological career.

The boat's name presumably comes from the gopher tortoise, *Gopherus polyphemus,* which even today can be found burrowing into Florida's mounds and middens. Moore may have intended the name as a metaphor for himself. The newspaper headline describing the steamer's launching notes that the "Gopher's Mission" was "To Hunt for Skulls and other Relics of the Mound Builders." The article goes on to say, "The lower deck will be fitted up with accommodations for the crew and also for the implements to be used in carrying off the excavations of the Indians mounds along the Florida streams. The upper deck will be fitted up with a private cabin and saloon for the use of Dr. Moore and the officers of the vessel" (*Evening Times Union,* August 28, 1895:1). These facilities also included housing for guests and assistants and a dormitory for locally hired African American field crews. Despite these conveniences, it was not long before Moore was privately complaining the vessel was too small (C. B. Moore to F. W. Putnam, letter, May 4, 1899, HUG1717.2.1, F. W. Putnam Papers, General Correspondence, Box 10, 1891–1900: G-M, Harvard University Archives).

Clarence Moore spent an aggregate of about ten years living aboard the *Gopher* and its chartered predecessors, which together saw hard service. The *Gopher* was constantly being grounded and often was hulled by snags. Aside from the continuous toll of rot and corrosion, machinery was always breaking down, the cabins often leaked, and from time to time the boat was infested with rats (Moore 1987, comments throughout). Yet Moore and his crew took these things in stride. He described his 1891 leisure trip on the St. Johns River as organized on the principle of "economy at any cost" (Moore 1987:Roll 6, Unnumbered notebook:3). Thirty years later he still was applying that principle.

When the field party was working, travel on the steamer went at a very slow pace. Stops were frequent, even in familiar territory, as Moore often inquired after the whereabouts of mounds and maintained his acquaintance with many people he had met. Certain ones, like the famously eccentric E. E. Ropes and "Captain" Mansfield, both on the St. Johns River, he seemed genuinely to enjoy visiting at every opportunity. Moore and his crew were enthusiastic fisherman and did not require strong persuasion to interrupt their digging or to tie up the steamer for a while if the fishing was good at some location, or else troll while under way. Further retarding their pace, they commonly encountered shoals and would tie up or anchor until the river or tide level rose. But more than anything else, the party stopped at every opportunity to take on wood to sustain the insatiable appetite of *Gopher*'s steam boiler. It may be no exaggeration to say that fully a third of their working time was devoted to this one activity (Moore 1987).

From late 1895 to at least 1918 Moore used the *Gopher* to reach and excavate not only mounds in Florida, where the bulk of his work was carried out, but also sites in Georgia, South Carolina, Alabama, Mississippi, Louisiana, Arkansas, Kentucky, and Tennessee, occasionally venturing into Illinois, Indiana, and Missouri. Moore literally steamed through most of the navigable Southeast.

In 1926, long after halting his fieldwork, Moore sold the thirty-one-year-old steamboat (Pearson et al. 2000:86). An anecdotal account by a south Florida resident describes the *Gopher* as "wrecked" in a 1926 hurricane (Tebeau 1955:60) that passed over the Gulf coast, but that was several months after Moore had sold the boat. Newspaper accounts following the hurricane indicate boats swamped by the storm were soon pumped and refloated. The aging *Gopher* may have been one such boat, for it continued to be used around Tampa Bay until at least 1927 (Pearson et al. 2000:86).

In the Field

Moore's archaeological modus operandi was much the same for three decades. Prior to many of the field seasons, contacts were made with local people and landowners to locate suitable mounds for investigation. This initial survey and contact often was done in the off-season by the *Gopher*'s captain, traveling in a small boat. A formal letter, on Academy of Natural Sciences of Philadelphia stationery, requesting permission to excavate followed this initial contact. Owners occasionally were paid to allow excavations. Landowners did not always grant Moore the permission he sought. Writing about his 1914–15 investigations on the Tennessee River, he noted: "We believe that the refusal of some of the owners to permit us to dig was not based on the firm belief in the presence of buried treasure in the mounds and sites, which prevails all over the South. . . . but on an exaggerated idea of the value of Indians objects" (Moore 1915a:181).

Moore began each season by traveling from Philadelphia to the location where his steamer was anchored. His principal assistant, Milo Miller, always accompanied him. Miller, who was a decade younger than Moore, was described by a colleague as Moore's "secretary, co-worker, physician, and friend who was to remain his inseparable companion to the end of his scientific work" (Wardle 1956). In fact, their association continued to the end of Moore's life. Dr. Miller was married and had two children. He was an austere personality who shared many intellectual as well as athletic interests with Clarence Moore. He had little interest in practicing medicine, though, noting to his daughter that he "would rather dig up dead people than look down [live] people's throats" (William F. Hitschler, personal communication 2000).

Milo Miller's contributions to Moore's work were great. He was salaried and also functioned as project anatomist—(today we would say biological anthropologist). He was responsible for much of the logistics that preceded each field season, and his hand also dominates Moore's field journals. In addition, Miller drew some site maps, including a famous one of the Moundville site in Alabama, and he helped shepherd Moore's many reports through the publication process.

Their field season generally began in late fall or early winter—November, December, or January—though on a few occasions it was earlier. Once the crew was hired and aboard the *Gopher* they departed for the field area selected for that year. A full day's work was usually about seven hours of digging, but very many days were spent tending to other matters, with the result that excavating often fell far short of seven hours in a day. When a mound was finished, the party steamed on to the next, at times traveling in the evening so that no time was taken away from excavation.

At the beginning of his work Moore put his photographic skills to use in making images of artifacts; later he hired the services of a professional photographer. But in one of the stranger quirks of this story, he was convinced that the camera was not an effective tool for archaeological field documentation. The background to this idea was that he believed it was impossible to combine small and large, or foreground and background, features in a photograph without obscuring or misrepresenting the scale and details of those features (C. B. Moore to F. W. Putnam, letter, January 6, 1898, HUG1717.2.1, F. W. Putnam Papers, General Correspondence, Box 10, 1891–1900: G-M, Harvard University Archives; Moore 1987:Roll 6, Unnumbered notebook:33). Consequently, very few site photos were made or published.

One can imagine the local interest Moore's entourage and its activities elicited when the *Gopher* steamed up to a site and excavations commenced. Contemporary newspaper accounts catch some of the excitement, but they also record much misinformation and rumors that locals must have traded among themselves. For instance, a *Florida Times Union* story dated May 20, 1895, headlined "From the Indians Mounds. Nearly One Thousand Skeletons have Been Unearthed at Mill Cove," notes that "Fully a thousand skeletons were unearthed," and "things considered of great value by Dr. Moore were discovered. The men say that they can't see what anybody wants with such things, that have been buried so long, when he can buy new ones for a fractional part of what it cost to dig down the mound. They may not understand, but the doctor does."

The mounds in the Mill Cove locality, including Shields Mound (8Du12, also known as the mound at Mill Cove), had first been the object of excavations by Moore in 1894. At the time, his investigations in Shields were limited because "the owner, a foreigner, alleg[ed] superstitious terrors on the part of his family" (Moore 1894a:204). In the 1895 field season, the period reported on in the story, Moore excavated a total of twenty-three mounds on the St. Johns River between Jacksonville and the river's mouth, including a second visit to the Shields Mound and work at other Mill Cove locality mounds. The number of human burials noted by Moore in his report on that season's work is about one hundred, not one thousand (Moore 1895).

In another extravagantly garbled case, in 1916, a great deal of excitement was stirred up in Philadelphia when reports came from Tennessee that Moore and three others in his party had been murdered (*Evening Bulletin,* April 20, 1916). Later, Moore's comment on all this was that it was more dangerous dodging cars while crossing Philadelphia's Broad Street than steaming around the South on his boat (*Evening Bulletin,* May 25, 1916).

Apart from the sensational and inaccurate news stories, Moore did experience a number of adventures as he steamed about the Southeast. In 1910 on the St. Francis River in Arkansas, the *Gopher* was forced to turn back when local residents—ardent segregationists—threatened the lives of the African American field crew if the expedition continued upstream (Moore 1910:256)—not the only time this happened. The spring before, in 1909, the *Gopher* was nearly sunk when the boat snagged a log on Bayou Bartholomew in northeast Louisiana (Pearson et al. 2000:82).

Fieldwork was not all hard work and peril, however. They rarely worked on Sundays, often took tea at four in the afternoon, and fished a lot. Moore's notes for April 1, 1905, when he was working at the famous Moundville site in Alabama, contain the following: "This P.M. the Gopher men wiped out the Moundvillians at base-ball—score 14–3" (Moore 1987:Roll 4, Notebook 25:266).

Field seasons usually continued until the end of April, when heat reached whatever location he was working and Moore headed back to Philadelphia to finish his reports. When a season was completed, several tens of sites would have been opened. Typically, though, half of these would have been investigated very briefly—for less than a day, often for only an hour or so. Sites worked for a few days represented a significant investment for Moore, and they only rarely spent weeks at sites like Moundville or Mount Royal.

Archaeologist or Grave-Digger?

Was Moore a competent archaeologist for his day, or was he, as one modern scholar has him (somewhat in jest), a "sophisticated grave digger" (Stoltman 1973:131). He certainly did many of the same things modern archaeologists do. He was well aware of the work of other archaeologists pertinent to his own research, and he did not hesitate to call on those colleagues and other specialists for information, including analyses that were beyond his own abilities. He readily shared his information with other archaeologists, and there is no record of Moore being envious or competitive when it came to the work of others.

Professional archaeologists of the day heaped praise, encouragement, and advice upon Moore: Putnam, Holmes, Hrdlička, Boas, Moorehead, McGee, and others all weighed in unambiguously. No less a personage than Smithsonian Institution archaeologist William H. Holmes, a longtime friend, used collections and information provided by Moore in studies of Florida and southeastern aboriginal pottery (Holmes 1894, 1903). In 1895 the *Archaeologist,* a journal with worldwide coverage edited by Warren K. Moorehead, reprinted excerpts from Moore's early articles and monographs on his St. Johns River excavations. On the first page was noted, "These articles will be compiled from Mr. Clarence B. Moore's publications, and will cover the most thorough and extensive work ever done among the sand mounds and shell heaps of Florida." Moore was elected a member of a number of American and European learned societies, including the American Antiquarian Society. Contemporary members of that society included such academic luminaries as Hiram Bingham, Franz Boas, Herbert E. Bolton, Frederick W. Hodge, Alfred L. Kroeber, and Samuel Eliot Morison. Even so, he stayed outside the circle of Philadelphia academic life. Moore also served for years on the boards of the Museum of the American Indian, Heye Foundation, and the American Anthropological Association.

Moore understood the principles of archaeological stratigraphy, and he generally looked for depositional patterns within the sites he excavated, recording his observations in his field notes. Moore did his homework; he read the literature quite thoroughly. He sought similarities between his sites and artifacts and those found elsewhere in the United States. His initial archaeological forays in Florida were guided by familiarity with published reports of Jeffries Wyman, Andrew E. Douglass, and S. T. Walker. In his initial report on the Mt. Royal site on the St. Johns River, he quotes from John and William Bartram's accounts of their eighteenth-century visits to that site, as well as that of Jeffries Wyman a century later. Moore also had read about the colonial French presence along that river in 1564–65 and even compared some of the metal artifacts he found to items worn by St. Johns-region Timucua Indians depicted in 1591 engravings done by the Flemish artist Theodor de Bry, based on French sources. In his later work in the interior Southeast, Moore cited widely from the growing body of archaeological literature.

In the margins of his notebooks from his early projects Moore occasionally jotted notes to himself, thoughts for future consideration, and reminders to seek additional information. "Contact with Europeans?" he queries in the field notes of one season at the Mt. Royal site in Florida. "Are shell heaps & sand mounds contemporary?" he asks in the draft report on the same site.

These are perceptive questions that Moore, in his early excavations in northeast Florida, sought to answer.

Perhaps the best work Moore did was his earliest: the excavations in shell middens and mounds in the St. Johns River drainage in the early 1890s. As James Stoltman has noted, "Moore was able to revisit all of Wyman's sites, to record forty-three additional sites, and to excavate on a larger scale than anyone before him. His work was surprisingly meticulous—and he proved to be a careful observer. He amassed further irrefutable evidence in support of Wyman's view that the middens were of human origin while, with *stratigraphic evidence,* he demonstrated that Wyman's opinion about the relative ages of the shell middens was indeed true" (1973:129; emphasis in original; also see Mitchem 1999:39–41).

Moore's archaeological work on the St. Johns River and, perhaps, his ability to mount field expeditions were apparently well respected by his Philadelphia peers. In 1895 the University of Pennsylvania Archaeology Department offered Moore the leadership of an expedition to a large south Florida concession they had acquired from the Disston Land Company (W. Pepper to C. B. Moore, letter, April 13, 1895, North America, Florida Explorations 1894–1903, Box 2, Folder 2, University of Pennsylvania Museum Archives). He turned it down (C. B. Moore to W. Pepper, April 21, 1895, North America, Florida Explorations 1894–1903, Box 2, Folder 2, University of Pennsylvania Museum Archives), however, probably because he was forever disinclined to associate himself with the university.

As Moore became increasingly enamored with excavating mounds, and later with cemeteries instead of middens (in the mid-1890s), he became more interested in finding high-quality ceramic vessels and other artifacts. The structure and exact nature of mounds sometimes was secondary to artifact recovery, and the phrase "nothing of value was found" appears far too often in his field notes and reports in reference to excavated mounds where no ceramic vessels were found.

This lessened concern for context is reflected both in his attitude toward field photography and in his field notes, which by the end of his career tend to be little more than a listing of artifacts found. The published monographs become heavily illustrated versions of the field notes. Some of the reports tend to be brief, but being able to match artifacts with sites provides modern archaeologists with important data. And the observations on context that can be teased from Moore's notes or reports sometimes can be gems for modern scholars.

In the field Moore worked too fast and tried to do too much, especially after he embarked on his mound excavations. This tendency apparently was not just an attempt to find as many high-quality pottery vessels as he could. In the field Moore witnessed firsthand the market for antiquities that already had developed in the Southeast by the turn of the century. Repeatedly in the introductions to his reports he makes reference to "treasure-seekers," "relic sellers," and "irresponsible" residents who looted sites for profit. He also railed against the artifact counterfeiters who were inflating a flourishing, early-twentieth-century antiquities market, contributing to the disappearance of archaeological sites. He viewed himself as a bona fide scientist, working under the auspices of the Academy of Natural Sciences of Philadelphia, whose excavations were saving the contents of mounds from those who would despoil the sites and sell their artifacts. The more mounds he could excavate, the more artifacts he could save. But by his own declaration, Moore saw to the "total demolition" of hundreds of mounds; "[w]here these mounds have not been leveled to the base, the fault has not been ours" (Moore 1894a:129).

His three excavations at the famous Crystal River site on Florida's Gulf coast (in 1903, 1906, and 1918) illustrate the rapidity with which they worked. Altogether, Moore spent 34 days at Crystal River with only about 25 days during which digging took place. In that relatively short time more than 40 intact (or nearly so) ceramic vessels and hundreds of other artifacts (including a variety of delicate copper and shell objects) were recovered. They also excavated the skeletal remains of at least 429 individuals. Clearing so many artifacts and burials with a crew of a dozen or even twice that many people in so short a time would be impossible utilizing modern standards of excavation.

Moore's crews must have literally ripped things out of the ground, hence the epithet that he was a "grave digger." Few detailed field drawings were kept, no grid system was used, and valuable measurements consequently were not always recorded, despite Moore's saying they were. What he meant was that he measured everything he perceived as important. Cursory observations regarding skeletal remains were made in the field presumably by Dr. Miller. Often these remains were forwarded to Ales Hrdlička for further study. But most of the skeletal material Moore thought was too fragmentary to collect and was reburied in the mounds while some was handed out as relics to local residents (*Florida Times Union,* November 28, 1895:7).

Moore deposited most of his excavated collections at the Academy of Natural Sciences of Philadelphia. He also gave artifacts to landowners and donated others to museums in Philadelphia, such as the Wagner Free Institute, as well as the Museum of the American Indian, Heye Foundation, in New York City, Robert S. Peabody Foundation for Archaeology (Andover), Field Museum (Chicago), Peabody Museum at Harvard University (Cambridge), and many other institutions, including historical societies. Artifacts even were sent to museums in Europe and South America (Moore 1907:363). Nearly all of the artifacts sent to such institutions are still available for study today. It is Moore's preservation of these collections, as well as the valuable monographs and articles he authored, that so warms the hearts of modern archaeologists toward him.

In 1929, however, the enormous Clarence B. Moore Collection at the Academy of Natural Sciences was lost to that institution. The general public impression, fueled largely by newspapers (*Public Ledger,* May 8, 1929:1, 10) and by Harriet Newell Wardle, assistant curator and sole staff of the Department of Archaeology at the Academy, was that the academy's newly appointed managing director, Charles Cadwalader, "young, ignorant of science, and eager to make his mark," decided the Moore collection had to go to make room for zoological collections. Cadwalader, said by Wardle to be "without museum experience," and employing extraordinary shenanigans, arranged the sale of the Academy collections to the Museum of the American Indian, Heye Foundation. Wardle further claimed that Moore, aged, retired from archaeology, and wintering in Florida at the time, was bitter about the transaction, but was forced to agree in order to prevent the collections being relegated to a leaky basement at the academy. After stimulating a brief flurry of interest in the matter at the University of Pennsylvania and in local papers, Wardle fell on her sword and resigned from the academy after a career there spanning more than thirty years (Wardle 1929). What she did not know was that Moore had engineered the entire transfer.

Archaeology and ethnology had never enjoyed a comfortable home at the Academy of Natural Sciences, a situation that had not eluded Clarence Moore's attention. An attempt in the 1890s to establish a department of Anthropology did not take hold. So long as the academy operated on a largely voluntary basis, with Moore paying all expenses associated with curation of his collections as well as personally watching over their placement, there was little problem. But he suspected that after his death the collection would receive careless treatment born of a lack of institutional interest. This impression was only reinforced in the mid-1920s when the academy, confronting a crisis about its role in the intellectual life of the community, explicitly refocused its mission and museum presentation on the natural sciences, and they hired the energetic Cadwalader to modernize the institution (Conn 1998:67–71). To Moore this only made the need to find a more reliable home for his collection—which he had deeded to the academy many years earlier—more acute.

In the spring of 1929, reasoning that Cadwalader would welcome an opportunity to dispose of the Moore collection, Moore contacted his friend George Heye to sound out his interest in the collection, thinking that Heye's Museum of the American Indian in New York would be just the secure kind of place for the Clarence Moore Collection. Of course, Heye, the avid collector of "every last dishcloth" fame, pounced on the opportunity. Heye and Moore agreed on a price ($10,000), and Moore told Heye to put the proposition to Cadwalader; Moore would support it when asked (G. G. Heye to C. Cadwalader, letter, March 25, 1929, No. 3, Box 121, #E, National Museum of the American Indian Archive [NMAIA]; C. B. Moore to G. G. Heye, telegram, April 6, 1929, Box OC121, No. 3, NMAIA; C. B. Moore to E. B. Morris, Sr., letter, April 21, 1929, Box OC121, No. 3, NMAIA). Cadwalader took the bait and the collection was quickly transferred to New York. Today it is part of the Smithsonian Institution's National Museum of the American Indian. Although Moore could not have foreseen this latter development, his intuition was right, and his decisive action insured the collection would be preserved.

But why did Moore not consider relocating the collection to the University of Pennsylvania Museum? Others asked this question at the time, and there is no recorded answer. We can infer, though, that this was owing, at least in part, to his continuing resentment over the role of university-affiliated persons in the 1890s conflicts with his mother.

Finale

Still going south to escape the cold Philadelphia winters, Clarence B. Moore had been under a doctor's care for only about six weeks when he died on March 24, 1936, at the age of eighty-four in St. Petersburg, Florida. Having been afflicted with various ailments for most of his adult life, he nevertheless outlived all of his immediate family. At his side to the end was his loyal friend, Milo Miller, who died six years later at his home in Atlantic City, New Jersey. As one of Miller's grandsons told us, "They'd been together a long while—and they seemed to have a consensus of purpose—

[they] were willing to endure a lot of inconveniences to do what they did" (C. W. Hitschler Jr., personal communication, 2000).

Ironically, Moore's death came at Mound Park Hospital, today the Bayfront Medical Center. Mound Park took its name from a large, conical shell mound on the hospital grounds, one of several nearby shell and sand mounds (Fuller 1972:10). Thirty-six years before his death, Moore had dug in two of those mounds.

Moore excavated several of the Southeast's most famous archaeological sites, including Crystal River and Mount Royal in Florida, Moundville in Alabama, and Indian Knoll in Kentucky. If he had worked only at those sites, he would have gained everlasting stardom in the annals of archaeology. But because he excavated so many sites—mainly mounds—he has achieved a unique status and importance far beyond that of any other individual in American archaeology. Would we have been better off if Moore had not dug Mt. Royal or Moundville? Certainly it would have been better if his excavations had been done to today's standards. But that could not have happened. The reality is that had Moore not dug and recorded what he did, we would have little or no information about many, if not most, of the sites he investigated. Treasure hunters eventually would have raided the sites, destroying them and looting their contents.

There are lessons here for all archaeologists. First, one's professional contributions rest not on excavation but on curation and publication. Clarence Bloomfield Moore certainly recognized that dictum more than a century ago. And second, despite the universal encouragement he received from the archaeological community, he might have done better to eschew excavation as his exclusive activity. Instead, he might have occasionally followed the example of his close friend, Frederic Ward Putnam, among whose many legacies was the acquisition and preservation of Serpent Mounds in Ohio in the mid-1880s (Brew 1966:17–18).

7

Lucy L. W. Wilson, Ph.D.

An Eastern Educator and the Southwestern Pueblos

Frances Joan Mathien

Her name is not well known to anthropologists in Philadelphia (see chapters in this volume) or in the American Southwest, yet Lucy Langdon Williams Wilson (fig. 7.1) conducted a study of southwestern archaeology and ethnology at a time when few women were responsible for such research projects (Babcock and Parezo 1988, Fowler 2000, Parezo 1993). For three years (1915–17) she excavated sites in the Otowi community located on the Pajarito Plateau of north-central New Mexico. She also collected ethnographic information from nearby San Ildefonso Pueblo that would aid in the interpretation of the artifacts she recovered. Her goal was to link prehistoric remains with historic Native American tribes in an exhibit at the Philadelphia Commercial Museum that would be designed to inform the public about modern and ancient Pueblo Indian lifeways.

Lucy Wilson was a well-known and dedicated Philadelphia educator who utilized many teaching techniques to reach students in her high school classrooms, as well as others who lived in the City of Brotherly Love. She had many scholarly interests. Among them was "ethnographic geography" (Shippen 1915); as part of her research into this broad topic, she traveled extensively, participated in ethnographic and archaeological projects, and prepared exhibits. Fieldwork in New Mexico, for example, provided material for an exhibit at the Philadelphia Commercial Museum.

This essay will provide some background for the early 1900s when archaeologists had only recently achieved status as professionals (Kehoe 1999) and when only a few women had obtained graduate degrees or carried out

7.1. Lucy Langdon Wilson. The Historical Society of Pennsylvania (HSP) from the Philadelphia Record Photo morgue (V:7 Surname Series)

fieldwork in the American Southwest (Babcock and Parezo 1988; Levine 1999:134; Rossiter 1982). It will introduce Lucy Wilson, her background, and achievements and document her participation in the collection of southwestern archaeological and ethnological data as the basis for the museum exhibit. Such an endeavor was unusual in a period when women's contributions to archaeology were usually through ancillary studies rather than as directors of such projects (Reyman 1999:214–17). Although there is much that is not known about her work in northern New Mexico, this review will provide a glimpse into the life of a dynamic and energetic woman who was a recognized Philadelphia leader during the early twentieth century.

Anthropology in the Early 1900s

Although a few women did contribute to anthropology in the late nineteenth and early twentieth centuries, their contributions often went unrecognized until recently (for example, Sara Yorke Stevenson [Danien and King, this volume]). As several essayists in Kehoe and Emmerichs (1999) point out, this period was one in which archaeology was in transition from the collection of artifacts to a science that addressed research questions. At the time only a few schools and museums were allied to promote scientific inquiry in this field. Among them were Harvard University and the Peabody Museum of American Archaeology and Ethnology, the University of Pennsylvania and its University Museum, and Columbia University and the American Museum of Natural History. Other institutions that supported anthropological research included the Bureau of American Ethnology, the Chicago Field Museum of Natural History, and, after 1908, the School of American Archaeology in Santa Fe. Many of the scholars who directed research projects were either trained in other fields or were self-taught. Kehoe (1999:5) indicates that in 1894 George Dorsey was the first to receive a Ph.D. in archaeology in the United States. She further points out (Kehoe 1999:8–9) that class interests and a need for patronage played a role in how and which projects were funded. She characterizes the archaeological leaders as WASP (white, Anglo-Saxon, Protestant) males who did not consider females equal or even capable of rational thought required by science.

For women who wanted an education or a career, the late nineteenth century provided some opportunities. Rossiter (1982) points out that there were a number of colleges that supported education for women but that the schooling was initially directed toward the creation of wives and mothers who could guide their children. During the 1880s and 1890s, advanced degrees became available at a few universities. In anthropology, women began to work in museums, accompany men into the field, or support field projects (Rossiter 1982:58). Levine (1999) discusses four women who contributed to Americanist anthropology at this time. Matilda Coxe Stevenson originally accompanied her husband and later conducted her own fieldwork in ethnology and archaeology in the Southwest. Her observations and reports provide a foundation for much of what was known about Puebloan life at the turn of the twentieth century. Alice Cunningham Fletcher's ethnological and archaeological research focused on the Midwest, and her support was essential in the establishment of the School of American Archaeology in Santa Fe. Mary Porter Tileston Hemenway sponsored the research of Frank Cushing and J. Walter Fewkes in the Southwest. Zelia Maria Magdalena Nuttall focused her studies on Mexican codices. Women were allowed to participate in some scholarly discussions and organizations but not others. This led Stevenson and others to establish the Women's Anthropological Society of Washington in 1885; it existed until 1899 when it merged with the (formerly men only) Anthropological Society of Washington (Levine 1999:135–38).

Because the American Southwest, with its numerous ruins and native peoples, had captivated the imagination of eastern scholars and the educated public after about 1850, many of the events that affected the course of American archaeology took place there. Fowler (1999) and Snead (2001) document the rivalries that arose between east and west, between trained anthropologists and local interest groups, between a younger generation schooled in anthropology and older investigators who learned through participation and applied lessons from other fields of study. By the second decade of the twentieth century, the studies of Nels C. Nelson and A. V. Kidder signaled the close of the descriptive period in which the direct historical approach was predominant (Willey and Sabloff 1980).

There were few women working in anthropology in the Southwest prior to 1920. Babcock and Parezo (1988) list three pioneers. One is Matilda Coxe Stevenson, previously noted. The second is Barbara Friere-Marreco Aitken (1879–1967), the first English female anthropologist. She obtained her degree from Oxford University in 1908 and a scholarship from the university before joining Edgar L. Hewett's School of American Archaeology summer field session in the Rito de los Frijoles in 1910. During that summer she studied Tewa ethnobiology with John P. Harrington. She continued ethnological fieldwork for several more years in the Southwest. Third was Elsie Clews Parsons (1875–1941), who received her Ph.D. in 1899 and taught at Columbia University. Parsons used her wealth to support her own work and that of others on comparative studies of historic Pueblo people. Parsons also founded the Southwest Society in 1918 to coordinate ethnological research. The remaining women listed by Babcock and Parezo (1988) were slightly younger and entered the Southwest after World War I, a decade after Lucy Wilson completed her work. Many had the backing of men who were unusual for their support of female students, including Byron Cummings of the University of Arizona and Edgar L. Hewett through both the School of American Archaeology and the University of New Mexico. Those in this later group include Florence Hawley Ellis, Marjorie Ferguson Lambert, Bertha P. Dutton, and Dorothy L. Keur. All four benefited from an association with Hewett (Mathien 1992).

In summary, the work of Lucy Wilson took place at the end of the descriptive era. The period during which she lived saw changes in education and opportunities for women in anthropology. Women generally were trained for careers as wives and mothers, as well as teachers. Advanced degrees in science were thought to be for males only. Anthropological projects tended to be sponsored by government agencies, museums, or wealthy patrons. Prior to 1920, those women who participated in southwestern archaeology—for example, Stevenson (Holmes 1916; Levine 1999) or Marietta Wetherill (Reyman 1999)—either tended to be assistants to their husbands or sponsors of expeditions. Most of their anthropological contributions were in ethnology.

Lucy Wilson's Life

Born in St. Albans, Vermont, on August 18, 1864, Lucy Langdon Williams was a member of the eighth generation of a family that contributed clergymen, statesmen, soldiers, and educators to the United States (Moore 1970–72:1).[1] By age thirteen she had graduated from the State Normal School. In 1879 she moved to Philadelphia where she enrolled as a second-year student in the Philadelphia High and Normal School for Girls. In 1884 she began teaching as a "temporary measure." In fact, teaching would absorb much of her time and energy throughout the rest of her life, but it would not limit her broader interests in the natural world and its inhabitants.

Between 1890 and 1897 Lucy Williams was a student at the University of Pennsylvania, where she studied biology and geography and obtained her doctorate in education. (She later took courses at Harvard University, Cornell University, and the University of Chicago.) In 1893 she married William Powell Wilson, then professor of anatomy and physiology of plants in the School of Biology at the University of Pennsylvania. In the same year, he founded and became director of the Philadelphia Commercial Museum (now the Philadelphia Civic Center). Lucy Wilson continued to teach and began to raise a family (Williams's daughter, Mildred, from a previous marriage and their son, David).

From 1892 through 1915 Lucy Wilson taught at the Philadelphia Normal School where she was head of the Department of Biology. In addition to her work at this school, in 1892 she organized the first Evening School for Women at the William Penn High School. She introduced some of the latest technological advances (for example, sewing machines, dressmaker forms, and Dictaphones). She also organized classes in salesmanship, developed progressive teaching methods, and applied laboratory methods to courses of study.

During this time Wilson's skills in creating exhibits were honed and recognized. She prepared for a number of expositions. In 1900 she was awarded a gold medal for a Philadelphia Normal School nature study exhibit that she personally sponsored at the International Exposition in Paris. The impact of this exhibit was so great that the City of Manchester underwrote it as a traveling exhibit in England, and the University of Aberdeen and the University of Dundee shared costs for sending it throughout Scotland.

Wilson was involved in ethnographic and archaeological research. Three newspaper articles (Shippen 1915, *Philadelphia Record* 1915, 1916) indicate that she focused on studies of man and nature. One article cited a long-term study of the effects of climate (*Boston Evening Transcript* 1916), another "ethnographic geography as represented by American origin" (Shippen 1915). Wilson had traveled to Bolivia, Ecuador, and Peru to study ruins and had published on her Peruvian studies by 1915. In 1913 Lucy and William Wilson visited seven Mayan ruins in the Yucatan. Previously, Lucy Wilson had visited Europe and Asia as well, observing archaeological techniques in Egypt.

From 1916 through 1934 Lucy Wilson was principal of the newly established South Philadelphia High School for Girls. During World War I, she and her husband were active in establishing the War Emergency High School to train women to fill positions needed to keep the country operating and supplying the front (*Philadelphia Inquirer* 1918a, 1918b). During this time she introduced to the Philadelphia school system the Dalton Plan that was used throughout New England for independent study. Lucy Wilson was also involved in studying educational systems worldwide. Research included trips to Russia in 1927 and 1932; based on her observations of the Russian family and educational system, she prepared numerous newspaper articles. During the 1920s, she was invited to lecture twice in Locarno, Switzerland, and four times in Nice, France; she delivered a series of addresses on educational theory at the University of Vienna, Austria; and she conducted guided educational programs in Santiago, Chile (Moore 1970–72; Who Was Who in America 1943).

Over the years, the Philadelphia newspapers carried several articles that document aspects of Wilson's pedagogical career. Wilson commented on many issues in women's education and there was considerable interest in her nomination for the superintendency of the Philadelphia schools, but the series of articles she wrote on her visits to Russia and her progressive ideas on education, which were based on her Russian observations, drew some accusations of communist leanings (Salerno, personal communication, 2000).

Lucy Wilson's outside interests were extensive. In addition to her career in education, she found time to write autobiographical books, edit *Everyday Manners* (1922) and *Education and Responsibility* (1926), and participate in several educational and civic organizations, as well as run a wholesale florist business (Moore 1970–72).

For her contributions to the field of education, Lucy Wilson earned the esteem of many of her colleagues, both in the United States and abroad. At the time of her retirement in 1934 she was selected as the person who had contributed the most to the City of Philadelphia in the preceding year and was awarded the distinguished citizen prize of $10,000 that had been established in 1921 by Edward W. Bok. She deposited this prize in a previously established endowment fund at the South Philadelphia High School for Girls to provide scholarships for deserving students; in 1934 this fund was the largest of its type available for such students.

After her retirement, Wilson remained active. She lectured on education at Temple University from 1934 through 1936. Much of her last three years was spent traveling. After suffering from a heart ailment for many months, she died in Lake Placid, New York, on September 3, 1937, at age seventy-three.

Wilson's Anthropological Project in New Mexico

Lucy Wilson became involved with the American Southwest through the Archaeological Institute of America (AIA). She and her husband were individual members, and the Philadelphia Commercial Museum was an institutional member. During the summer of 1914 (when she was forty-nine and he was sixty-nine), the Wilsons participated in an AIA School of American

Archaeology tour of New Mexico under the direction of Edgar L. Hewett. They visited several prehistoric and historic pueblos on the Pajarito Plateau and participated in limited excavations at one site. Among the sites visited was Otowi Pueblo. During this time they met Wesley Bradfield, an archaeologist who worked for Hewett. Correspondence in the Hewett files at the Museum of New Mexico suggests that the Wilsons thoroughly enjoyed their trip even though William Wilson suffered from angina attacks (L. Wilson to Hewett, January 24, 1915, Laboratory of Anthropology, LA 169 file). Lucy Wilson quickly realized she could possibly fulfill a dream she had had since reading about Schliemann's excavations at Troy and, at the same time, perform a service to eastern Americans (Wilson 1916b:29).

In a January 5, 1915, letter to Hewett (Laboratory of Anthropology Files, LA 169), William Wilson outlined the research goals of their project, which included excavating a large pueblo and interpreting the daily life of its inhabitants through studies at a neighboring pueblo.

> So far as Otowi is concerned, I should want to make a map not only of the pueblo, the burial mounds, reservoirs, etc. but of the surrounding country, particularly the pueblos connected with it by trails, and, of course, of past and present water supplies. This I should supplement with photos, telephotos and enlargements, with herbarium specimens, and with such material as I might have the good fortune to dig up, so as to give our people in Philadelphia, especially in the schools, some adequate idea of these most interesting prehistoric people. I want to make this exhibit as full as possible. But even at the best, it will need the illumination of a modern pueblo. I, therefore, propose to make as complete an exhibit of the present arts and occupations and people of San Ildefonso as the means at my disposal will permit.

At this time William Wilson also asked Hewett for advice because he had previously worked at Otowi and had made friends with a number of the people from San Ildefonso, some of whom had been his field hands during prior excavations on the Pajarito Plateau and who would return to work with Lucy Wilson.

Otowi

Otowi Pueblo (LA 169) is located about five miles west of the Rio Grande and six miles west of San Ildefonso Pueblo. In an open area that is approximately 1 mile square are two east-west trending ridges. To the north and south are high mesas; the northern one is known as Otowi Mesa. Two parallel canyons run through the glade, one on each side of the ridge that Otowi occupies. In several directions are cliff dwellings, which are carved into tentlike volcanic formations that have naturally formed cavities as well. The most unique set of tent rocks lies about half a mile upstream from the main Otowi pueblo (Hewett 1906:18–20, 1938:47).

Although Bandelier (1892:78) mentioned Otowi in his initial description of the ruins of the American Southwest, it was Hewett who first provided a good description of the site and who carried out the first excavations in the early 1900s. In addition to "Big Otowi," Hewett (1906:18) indicated the presence of a smaller pueblo (now known as Little Otowi, LA 32) and seven smaller sites located about three hundred yards across the southern arroyo on a parallel ridge.

To the people of San Ildefonso, Otowi was known as Potsui'i, "the pueblo ruin at the gap where the water sinks" (Harrington 1916:271). Hewett, who had become a confidant of "Wayima," a headman at San Ildefonso learned from him of an oral tradition that prior to moving to Perage (the ancestral village to San Ildefonso) on the west side of the Rio Grande, they had lived in smaller settlements on the adjacent mesas, including Otowi. Their reason for moving down into the valley was a lack of water. Hewett (1938:48) indicates that some other clans at Powhoge claimed to be descendants of occupants of Tsankawi (another Pueblo ruin in the area). In 1906 Hewett suggested that the people from San Ildefonso had lived along the Rio Grande for six to eight centuries. He also thought people had moved from Otowi to Perage within a few generations.

Hewett (1904:642) had prepared a plan map of Otowi. He initially identified ten kivas, two of which were enclosed in Sections A and E. Section A was thought to be a single story high, Section B probably had two stories, Sections C and D might have been three stories high, and Section E may have had some fourth-story rooms. All but Section E were connected by walls. Hewett (1904:643) estimated 450 ground-floor rooms. Although he could not date this large house, Hewett (1904:643) indicated Section C would have been built first; its inhabitants would have been joined by other groups from the local small sites who, within the span of a few years, would have constructed the remaining sections. A reservoir to the southeast of Section A and two trash mounds comprised the remaining features at this major site.

Little Otowi was described as a rectangular building with one kiva in the court. The stones used in construction were smaller and the laying of rocks was cruder than the masonry at Big Otowi (Hewett 1906:20, 1938:47). This was probably an earlier building, which, because of its size, may have been the center for the seven other small houses on the same ridge.

By 1904 Hewett had excavated in a number of the smaller pueblos on the Pajarito Plateau and realized that they were earlier than Otowi and the other large sites there. This conclusion was based on pottery types and "traditionary evidence" (Hewett 1904:647). At these small sites, there were few kivas, and no burials or cemeteries had been found. In 1905 Hewett recovered parts of more than 150 burials from the two trash middens at Otowi. Many were disturbed by later interments. Burial types were in pits in a rock floor, secondary burials in ceramic vessels, and were located throughout the mound itself. Approximately 20 percent were infants (Hewett 1938:130–32).

Thus, by the time the Wilsons participated in Hewett's tour of the Pajarito, Hewett had already collected basic information about the valley in which Otowi and the other smaller sites were located. As a result of his 1905 excavation, an artifact collection that included pottery, bone awls, polishing stones, and ceremonial pipes or "cloudblowers" was already curated at the U.S. National Museum. The research that was proposed would increase knowledge at a third large pueblo on the Pajarito, in an area intermediate to Hewett's prior excavations at Puye (ancestral site to the Tewa people of Santa Clara Pueblo) and Tyuonyi (ancestral site to the Keresan people of Cochiti Pueblo) (Hewett 1938:102).

Wilson's Ethnographic Studies and Archaeological Excavations

In the spring of 1915, planning for the project was under way. During the initial year of their work in 1915, the Wilsons relied on Hewett to obtain their excavation permit. The School of American Archaeology held the permit during the first year as a concession to the Philadelphia Commercial Museum. In the following years, this permit was issued directly to the Philadelphia Commercial Museum; it covered the excavations and permission to retain the collections. Hewett also provided advice about camp locations, personnel needs, and costs. The Wilsons needed a project manager who was familiar with excavations, supplies, and camp needs. Lucy Wilson (who was responsible for the field work) requested the assistance of Wesley Bradfield,

TABLE 7.1.

List of Native Americans who worked with Lucy Wilson*

Name of Worker	Glass Slide No.
Museum of New Mexico Photo Archive	
Santiago	82811, 82915, 82921
Facundo	82873
Atilano	82871
Crescencio, Anna and Santiago	81998
Agapito	82921
Museum of New Mexico, Laboratory of Anthropology	
Crescencio	121, 132, 133, 136, 153,
Miguel	124, 101
Santiago	unnumbered, Box 53, 184, 76
Dronsain	unnumbered, Box 53
Alfonso	101

*Taken from annotations on photographs in the Museum of New Mexico Photo Archive and glass slides in the Hewett Collection in the Laboratory of Anthropology.

who became the field officer in charge. Assisting Bradfield was Captain C. J. Troutman, who was affiliated with the New Mexico Military Institute in Roswell, New Mexico (8th Annual Report of the School of American Archaeology to the AIA, 1915, p. 43) and acted as cook for the expedition.[2] Wilson also hired a number of men from San Ildefonso (table 7.1) who had worked with the School of American Archaeology in previous years (Shippen 1915).

Lucy Wilson planned to spend one week obtaining information from the inhabitants of San Ildefonso and five weeks on excavations at Otowi. By July 27 she was in New Mexico. Archaeological work during the summer included a general survey of the region and excavations in several areas (a small burial mound, a few rooms in each of the large houses, thirty-six rooms in the E-shaped house, and trenching east of the reservoir) at Big Otowi (fig. 7.2). In a passageway between the outside wall of the North House and a heavy stone wall was a basket burial with pipes, lightning sticks, and prayer sticks. As a result, this material was interpreted as a cacique's burial (see the published description in Wilson [1916b:34–35; 1918a:316; 1918b:292–93]). By August 23, 1915, a scale map of the ground plan of Otowi Pueblo (Wilson 1916b:28) had been drafted by J. Percy Adams, who worked for Hewett. Numerous glass slides, as well as a collection of black and white photographs, were produced. Hewett requested a short summary of that summer's work for publication in *El Palacio* (Wilson 1916b). Lucy Wilson prepared a short report on Commercial Museum letterhead for the Department of Agriculture (on whose land Otowi was located), a copy of which was also transmitted to the Department of the Interior by Hewett. The collections of pottery, bone, stone, fiber, wood, and botanical remains were shipped east and were undergoing curation at the Philadelphia Commercial Museum.

In February 1916 Bandelier National Monument was established; it included an area to the northeast known as the Otowi Section. In June 1916 the Department of the Interior transferred the excavation permit directly to the Philadelphia Commercial Museum. This new permit covered a three-year term (1916–18) and both Lucy and William Wilson were listed as supervisors of the project. There was one stipulation, that "all uncovered walls be repaired to prevent further disintegration in the near future" (Wilson used the excavated materials as backfill in the rooms after they were emptied [*Philadelphia Record,* August 31, 1916]).

During the 1916 season, Rudolph W. Schiele from the Philadelphia Commercial Museum assisted the Wilsons (Anonymous 1916:84), acting in a position similar to that of Wesley Bradfield during the previous year. Again, men from San Ildefonso Pueblo assisted with the excavations.

> Our Indians are Tewas from the village of San Ildefonso. Usually it is difficult to get them during the harvest season. But this year, the grasshoppers destroyed their corn, alfalfa and gardens so that only the wheat

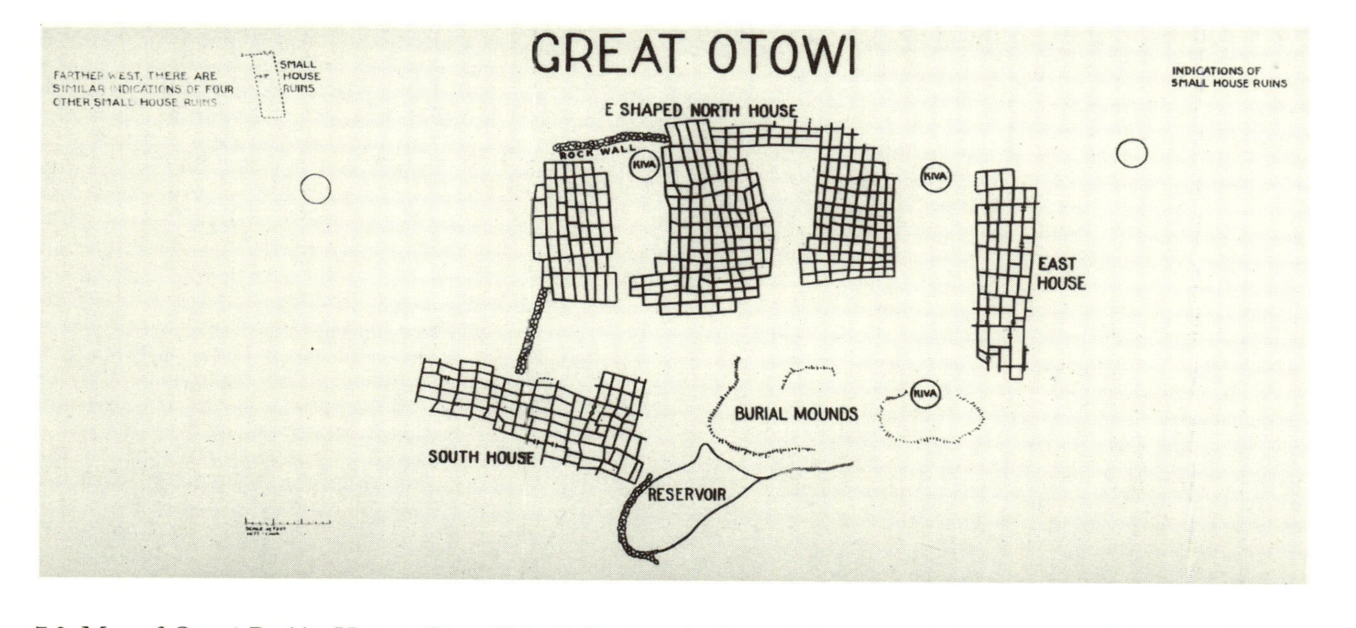

7.2. Map of Otowi Pueblo. Hewett Glass Slide Collection, Archives, Laboratory of Anthropology, Museum of Indian Arts and Culture, Santa Fe. (Box 53, Neg. #115)

> could be reaped; this disaster meant, too, that it was the more imperative for them to earn money to buy corn. In consequence, they have taken turns at harvesting and I have had ten or eleven Indians with me at all times, but only one of them has been with me every day. They receive $1.50 for an eight-hour day, from 8 to 12 A.M. and 1 to 5 P.M. I suggested trying 7 to 12 and 3 to 6 or 4 to 7 if the preferred. After consultation they decided to keep their old hours, saying that they felt a little lazy in the morning and that they liked to smoke. As a matter of fact, they waken with the sun at 5, sing, pray and make breakfast until about 6:30, and then subside into silence, smoking cigarettes until the call comes for work.
>
> They enjoy the work very much, especially when anything of importance is found. Then those near by suspend work to watch the delicate process of uncovering and taking out the treasure. Often after supper they stroll over to my tent to look over the finds and to comment upon them. Sometimes they copy the designs on the pottery. This year two modern bowls were brought to us, each with an adaptation of a design uncovered last year. Of course I bought them [*Boston Evening Transcript* 1916].

These quotations from Lucy Wilson indicate continuity in hiring some of the same workers for both years. Her descriptions of the materials recovered and the interpretations of these finds (Anonymous 1916, *Boston Evening Transcript* 1916, Wilson 1916a) also suggest an exchange of information between Wilson and her San Ildefonso workmen was ongoing during the excavations. Her published reports reflect the use of ethnographic analogy for interpretive purposes.

During 1916 a second general survey of the region was conducted. Excavations in 165 rooms in Otowi Pueblo, 16 rooms in smaller pueblos to the south, 4 talus rooms in cliff dwellings to the north, and 2 caves in the cliff houses were carried out. Photographs were taken. Skeletal remains and numerous artifacts were recovered. A summary of the summer's work was published in *El Palacio* (Anonymous 1916). Another article (Wilson 1916a) was devoted to the description of a clay figure found in an unusual room in Otowi Pueblo.

On April 18, 1917, the Wilsons transmitted a report on that summer's work to Frederick Webb Hodge of the Bureau of American Ethnology. In their cover letter they indicated that two individuals at the Commercial Museum were mending pottery and curating objects; the collection was open to inspection by anyone from the Smithsonian. They also indicated that several errors in the previous year's description of Otowi had been corrected, and they outlined the activities completed.

A third and final year of work took place in 1917. Lucy Wilson still relied on help from Hewett. In an October 23, 1916, letter, she requested assistance locating a young man who could cook, pack, and keep the camp in order from July 1 to August 15, 1917. Salary would be $100, but if he could survey as well, it would be $120. John Walter, son of Paul F. Walter of SAR, acted as cook, alcalde, and survey/mapper (see also Anonymous 1917).

In 1917 excavations were completed at a number of small sites and in Little Otowi and Big Otowi. A new corrected map of Otowi Pueblo was drawn, and all pueblos in the area were located and mapped. Several cavate sites were explored. The most interesting discovery that year was a painting of a mountain lion in a large square ceremonial room located near the middle of the East House at Otowi (Wilson 1917a, 1917b); the mural was recorded by Crescencio Martinez (fig. 7.3), a well-known watercolor artist.[3]

Although the permit issued to the Philadelphia Commercial Museum covered one additional year, no work was done in 1918. The United States had entered World War I; back in Philadelphia, Lucy Wilson turned her energies toward the war effort. The War Emergency High School for Girls opened on July 1, 1918, and ran through August 30, 1918. Wilson was one of the main organizers of the curriculum (*Philadelphia Inquirer* 1918a, 1918b). In an August 9, 1918, letter to Hewett, Wilson indicates she gave a forty-hour course at the school (Hewett Papers, Box 55).

Lucy Wilson's accomplishments during her three years at Otowi were summarized in several brief articles (Anonymous 1916–17; Wilson 1916b, 1917a, 1918a, 1918b). Excavations had been carried out at Big Otowi where the East and South houses were estimated to be only one story in height; in contrast the North house probably had a terraced second story and possibly a few third-story rooms (Wilson 1918b:293–94). (These houses are not quite as high as Hewett had estimated; he thought one section might have had some fourth-story rooms.) At Little Otowi, two buildings were excavated and some work was done in a nearby community of about the same size. In addition, at least seventeen other small, but earlier, sites were identified on the same ridge. On a lower ridge three other small-house ruins were noted (Wilson 1918b:292). Some rooms in cavate sites also were examined. Unfortunately no final report ever appeared, even though Lucy Wilson did prepare a two-part paper summarizing the project (see below).

7.3. Crescensio Martinez drawing mountain lion. Hewett Glass Slide Collection, Archives, Laboratory of Anthropology, Museum of Indian Arts and Culture, Santa Fe. (Neg. #153, Box 54)

The Exhibit

Wilson and the museum staff prepared an exhibit based on her work (Anonymous 1916–17). It included material on the environment, food, clothing, shelter, transportation and commerce, art, history, and religion. An August 9, 1918, letter from Wilson to Hewett indicated that she needed to spend a few additional weeks at Otowi to better develop the exhibit. She had spent the previous Christmas holidays (when Hewett was in Philadelphia) working with her slides and drawings and had hoped to show Hewett the Otowi exhibit personally and obtain comments (especially on the "awanyu" series) from him, but he left before that occurred. I have been unable to find evidence that she came to New Mexico later that summer or additional documentation of the exhibit. There is no record that indicates Lucy Wilson lectured on her archaeological work in the list of "Free Illustrated Lectures" sponsored by the Commercial Museum between 1915 and 1922 in the materials curated at the Historical Society of Pennsylvania (Wn 823, v. 1–2; Salerno, personal communication, 2000). She did, however, give a presentation of her three years' work at Otowi at the Fourteenth Annual Meeting of the American Association of Museums in 1919 (Anonymous 1919). Numerous slides in the Hewett Collection at the Museum of New Mexico illustrate the types of artifacts that were recovered. It is easy to imagine what such objects would have contributed to the knowledge of Pueblo culture to the people of Philadelphia.

The Collections

The collections include nearly a hundred whole pieces of pottery in addition to a very large and varied collection of shards; a case full of ceremonial objects, including a specifically fine collection of ceremonial cloud blowers; musical instruments and other bone artifacts; and stone utensils and weapons. "Its unique features are (1) a basket burial of a man of some importance, presumably a priest; (2) a small clay idol with turquoise eyes and heart—black with soot, and (3) a photograph of a wall fresco, in color, of a mountain lion" [Anonymous 1916–17:4].

As the Wilsons and Hewett had agreed at the time of the excavations at Otowi (Lucy Wilson to Hewett, June 19, 1933, Hewett Papers Box 35), the artifacts and photographs from the expedition were returned to the Museum of New Mexico in 1938, after both William and Lucy Wilson died (he in 1927; she in 1937), but before Hewett passed away. For ten years prior, the Laboratory of Anthropology had corresponded with the Philadelphia Commercial Museum in an attempt to obtain the collection. In 1937 Hewett learned that the Philadelphia Commercial Museum no longer wished to curate the collection and contacted Charles R. Toothaker, the curator. Toothaker did not realize there were two museums in Santa Fe; therefore, he packed the collection and shipped it to Hewett at the Museum of New Mexico in early 1938. (See correspondence files at the Laboratory of Anthropology, Museum of Indian Arts and Culture, for LA 169 in 1938; today the Laboratory of Anthropology is a part of the Museum of New Mexico).

Although no notes accompanied the collection, as it was being curated, Marjorie Ferguson Tichy (now Lambert) was able to elaborate on two finds. First, a number of gaming pieces were cataloged; they possibly represent dice

(Tichy 1941). The second, the painted mountain lion fresco, was placed into the larger context of painted murals of the area (Tichy 1947). The number of items in the collection at the Museum of New Mexico compares well with the published numbers reported by Lucy Wilson (Anonymous 1916–17:4). A note in the SAR annual report for 1939 indicates that the collection included 139 vessels, miscellaneous bone, shell, and stone artifacts, several unrestored vessels and some skeletal material, plus an excellent set of several hundred slides, duplicate negatives, and a few photographic enlargements. In 1989 Michael Stanislawski inventoried the collection for the National Park Service; more than 387 lithic items, 190 bone items, and 294 ceramic items were among the pieces counted. The ceramics included 94 bowls and 40 jars, numbers that compare well with earlier reports of the materials in the Philadelphia Commercial Museum collections.

It is unfortunate that there are no field notes or detailed reports of Lucy Wilson's work or the resulting exhibit. Correspondence from Wilson to Hewett on January 18, 1933 (Hewett Papers, Box 35) indicates that she did prepare a two-part report with illustrations for *Scientific Quarterly*. The first of two parts was lost in the mail. Although urged by Dr. Cottell (presumably the editor), Wilson never had time to rewrite that part because all of her energy went into various phases of education. A letter in the files at Bandelier National Monument dated May 23, 1954, to Thomas B. Onstott from Dr. David Hawxhurst Wilson (son of William and Lucy) reiterates that a draft of an article for *Scientific Monthly* in 1919 was lost in the mail, that no duplicate copy had been kept, and that his mother's notes were not full enough for her to rewrite this contribution quickly. David Wilson did not indicate if her notes were extant in 1953. Other correspondence between Onstott and Marie L. Morley, secretary to the Curator of the Philadelphia Commercial Museum in 1953, indicate that no notes or materials were ever turned over to that institution. Another letter in the files at Bandelier National Monument (from Ed Grusheski, Curator of Education, to John D. Hunter, Superintendent of Bandelier National Monument, October 26, 1981) suggests that if there were such notes at the museum, they probably were destroyed in a fire in the early fifties. As a result, it is difficult to assess the work that Lucy Wilson conducted at the Otowi community and to establish artifact provenience.

Comments

Lucy Wilson was a product of her time. As a member of a family with deep roots in the United States, she chose a career of service in the field of education. Her enrollment in the University of Pennsylvania in 1890 came four years after the university established its graduate program (Cheyney 1940: 296–308). Her Ph.D. was in education, an acceptable field of study for women at that time (Levine 1999:134). That she was committed to further women's education is apparent in her establishment of several schools (the Evening School for Women at William Penn High School in 1892 and the War Emergency High School in 1918) and the scholarship fund at the South Philadelphia High School for Girls that furthered women's education, as well as her 1934 contribution of her $10,000 award. She excelled in her profession and was recognized in that field.

Wilson's interest in anthropology had a long history; her travels and observations allowed her to recognize an opportunity to incorporate fieldwork in archaeology and ethnology into her educational program. She was fortunate to have been a member of the AIA and to have a husband who was director of an affiliated museum that could provide the needed sponsorship for the venture as well as a facility to curate and present the exhibition. She was fortunate too to have known Hewett, who, like Wilson, was trained as an educator, was curious about all aspects of the study of man, and was used to interaction with educated women, such as Alice Fletcher, who helped him establish the School of American Archaeology. Hewett easily incorporated knowledgeable individuals into archaeological programs on the Pajarito Plateau (Snead 2001). Thus, for this project, Wilson was not strongly affected by the professionalization in the sciences, including anthropology, that had been taking place since the 1890s. That she had sufficient means to pursue her interests is apparent by her ability to finance the high school poster at the 1900 Exposition in Paris, her travels, and the expeditions to Otowi.

Today we may be dissatisfied with the missing documentation of the fieldwork, the exhibit, and the unpublished results from Wilson's Otowi Project. Yet, measured against standards extant in 1918 and the products from contemporary projects, Lucy Wilson did as well as many of the professional archaeologists working in the field at that time. Her work reflects the then standard descriptive approach and use of ethnographic analogy.

Wilson's methods were also similar to that of most others. An article in the *Boston Evening Transcript* (1916) describing the work at Otowi indicates that two men were assigned to each excavated room; they placed all objects on the edge of their place of discovery, and Wilson noted the depths of these finds. After the adobe floors were reached, they were sounded to determine whether there were hollow or soft places beneath or they were on bedrock. In many rooms the most valuable finds were recovered beneath the floors. After her first year, she reported, "We have a record of the depths at which the skeletons and most of the pottery were found. Another year, however, we shall come prepared with large copies of the map, on which all measurements and finds may be promptly and easily recorded. There is no good reason why the methods of classical archaeology should not be followed in these excavations. I suppose that the fact that the finds came so thick and fast and that their intrinsic value is relatively so small, accounts for the fewer measurements and the less constant use of the sieve" (Wilson 1916b:32–33). This quote, plus the description of her methods that appeared in the *Boston Evening Transcript,* suggest she observed much during her earlier visits to archaeological sites and was able to use that knowledge, and any assistance she would have received from Bradfield and Schiele, in her excavations on the Pajarito Plateau.

Although she still depended on men for technical advice, Wilson was definitely the director and organizer of her own project. William Wilson's name and museum were on official correspondence and reports, but it was Lucy Wilson who traveled to New Mexico, carried out the ethnological studies, recorded the archaeological finds, took responsibility for the exhibit, and prepared the articles published solely under her name in several respectable journals. For the latter, however, she did not use her full name; authorship appears as L. L. W. Wilson. This use of initials may have been intentional. H. Marie Wormington always used her initials because she had been told that "nobody would read a book written by a woman" (Parezo 1993:5).

Lucy Wilson definitely accomplished her goals. Using ethnographic analogy, she was able to interpret the materials she excavated. There was an exhibit at the Philadelphia Commercial Museum, and the artifacts and photographs from the excavations were preserved in a curatorial facility. It would be over a decade before female archaeologists would direct excavations sponsored by Hewett in Chaco Canyon (Joiner 1992:56–58, Mathien 1992) or by A. V. Kidder at Tecolote, New Mexico (Preucel and Chesson 1994), let alone publish and present their results to the general public.

Acknowledgments. I would like to thank those individuals who helped me locate information on Lucy L. W. Wilson, William P. Wilson, and their anthropological project at Otowi. Eleanor King provided several leads that aided in locating archival material on Lucy Wilson. James E. Snead kindly provided copies of permits issued to the Philadelphia Commercial Museum for the excavations at Otowi, as well as references to other background material for this chapter. Their assistance is greatly appreciated. The information in this chapter, however, reflects my own understanding of Lucy Wilson's career and research.

Notes

1. This summary of Lucy Wilson's career is based on several sources: a listing in *Who Was Who in America* (1943), published references on archaeological excavations, and material culled by Abigail Salerno, research services associate at the Historical Society of Pennsylvania. Among the materials on file at that institution is a brief report by Ellen Moore, copies of several newspaper articles from the *Philadelphia Record* and the *Boston Evening Transcript*, and photographs used in local newspaper articles. Some citations provided in this paper are given in the format suggested in Salerno's letter dated July 5, 2000. All citations for Salerno, personal communication, refer to information included in her letter. Additionally, letters on file in the Hewett Papers at the Museum of New Mexico History Archive and in the Laboratory of Anthropology site report for Otowi Pueblo (LA 169) provided information on the relationship between the Wilsons and Hewett's School of American Research. By correlating the information from all these sources, it is possible to reconstruct a picture of Lucy Wilson's pedagogical career and her anthropological work in New Mexico. Throughout this section, references are cited only for major points; readers are referred to the Historical Society of Pennsylvania and the Museum of New Mexico for details.

2. A letter from Troutman to Hewett dated July 19, 1926, filed at the Museum of New Mexico, Hewett Files, indicates he was the cook for the expedition.

3. Crescencio Martinez was one of the men from San Ildefonso who worked for Lucy Wilson at Otowi; he was also recognized as a talented artist (Brody 1997). During the summer excavations, he was assigned the task of drawing the mural of the mountain lion as part of the permanent record of the summer's work. In 1918, "Crescencio Martinez . . . brought to our attention a number of water color paintings executed by him, illustrating the costumed figures in certain ceremonies performed by his people. Struck by the artistic merit of these as well as by their ethnological value, we engaged him to produce, in water color, pictures of all the characters that appear in the summer and winter ceremonies. These he finished just before his untimely

death" (Hewett 1938:126). Martinez's work was supported by the School of American Archaeology (Hewett 1918:67–69). Shortly after learning of his death, Lucy Wilson wrote to Hewett on August 9, 1919 (Hewett Papers, Box 55), praising Martinez's drawings and expressing her admiration of his talent. Wilson thought that he had been a fine spiritual leader of San Ildefonso (see also Anonymous 1918).

8

The Second Largest City in the English-Speaking World

John L. Cotter and the Historical Archaeology of Philadelphia, 1960–1999

Robert L. Schuyler

In popular American history it is sometimes claimed that by the middle of the eighteenth century or by the outbreak of the American Revolution, Philadelphia was "the second largest city in the English-speaking world." Such a demographic privileging is almost certainly wrong if cities on a scale of a Dublin or an Edinburgh are noted. This claim also draws attention away from a more interesting fact: eighteenth-century Philadelphia was indeed one of the more cosmopolitan cities in the English world in spite of its late founding date (1682) and its distance from Europe.

A similar oversimplification is sometimes assumed for the history of historical archaeology in regard to one of Philadelphia's leading scholarly citizens—John Lambert Cotter (1911–99). Cotter is viewed by many as an original founder of Americanist historical archaeology and is highlighted as the excavator (or one of the excavators) of Jamestown. Both attributes are historically misleading and detract from Cotter's true significance and chronological positioning within the history of the discipline.

Historical archaeology was initially established in North America as a professional endeavor during the 1930s and 1940s, the crucial event, especially in the United States, being the Great Depression. Numbers of excavators, all trained in North American prehistory, moved onto historic sites across the continent after 1933. From California missions to forts in the

8.1. John L. Cotter (1911–1999) Photo courtesy of the University of Pennsylvania Museum of Archaeology and Anthropology.

Great Lakes region to early sites in the Southeast, these archaeologists added the exploration of European settlements to the already existing theme of contact-Native American sites to create a new discipline. The central figure within this group was Jean Carl Harrington (1902–98), known to his colleagues as J. C. or "Pinky" Harrington (Miller 1998). He was the founder of the field in the United States, with Kenneth Kidd (1906–94) being his counterpart in Canada. Harrington was trained at the University of Chicago in prehistoric archaeology; however, in 1936 when he took over the already existing National Park Service-Civilian Conservation Corps (NPS-CCC)

Jamestown Project he moved almost exclusively into historic sites studies. Jamestown was excavated for six years (1936–41) with Harrington returning in 1949 to explore the glasshouse location.

Harrington and his colleagues developed basic field approaches for colonial sites, and he initiated, along with scholars like Arthur Woodward, the study of historic assemblages. After Jamestown he went on to work on Raleigh's sixteenth-century colony on the coast of North Carolina (1947–50) and Fort Necessity in Pennsylvania (1952–53), with work at nineteenth-century sites coming later in his career. This truly pioneering generation published the first site reports, established basic methodologies, and tentatively came to understand historic artifacts and architecture. Equally important were Harrington's first attempts to define and survey work in the new field, especially his 1952 article ("Historic Site Archaeology in the United States") in Griffin's volume dedicated to Fay-Cooper Cole, and to issue well placed programmatic statements for the field—for example, his 1953 "Archeology and Local History" in the *Bulletins of the American Association for State and Local History* and his 1955 "Archeology as an Auxiliary Science to American History" in the *American Anthropologist* (Harrington 1952, 1953, 1955).

By the mid-1950s historical archaeology was a visible and discrete topic within general archaeology with a growing number of practitioners scattered across North America. These archaeologists were starting to interact with one another and to set themselves off from their colleagues in prehistoric studies.

John Cotter had a two-decade professional history that parallels these developments between 1934 and 1954, but this career was in North American prehistory with no significant ties to historical archaeology. Cotter's fieldwork focused on Paleoindian studies, more extensively on southeastern mounds and administratively on the prehistoric Southwest. He was not a member of the pioneer founding generation of historical archaeologists, at least not a member of its first cohort (Roberts 1999a).

In 1954 Cotter took field direction of a second major project at Jamestown as that site moved toward its 350th anniversary (1607–1957). Joined by Edward Jelks, Louis Caywood, Joel Shiner, and Paul Hudson, Cotter engaged in three seasons of extensive testing and excavations resulting in his 1958 "Archaeological Excavations at Jamestown, Virginia," (*Archaeological Research Series* No. 4, National Park Service, Washington, D.C.), which incorporated Harrington's earlier manuscript reports and cartographic work. Since 1958, John Cotter's career is usually viewed with Jamestown as the backdrop. Such an image is factually acceptable, but like Philadelphia's "second largest" designation, it greatly underestimates and misplaces his primary contributions. He did not continue to work on seventeenth-century sites, he had no prior or post-Jamestown experience on such early sites or assemblages, and his well-done, potentially landmark report was blunted in its impact by the refusal of Jamestown to carry it in the gift shop and by the governmental early remaindering of the handsome hardbound volume into the shredder (Roberts 1999b). It was not Jamestown but rather Philadelphia that was central to Cotter's career in historical archaeology, a career that postdates 1960.

John Lambert Cotter was a member, indeed perhaps the central figure, of a *second* cohort of researchers who built historical archaeology from the mid-1950s across the succeeding decade. If 1936 is the key date for the first phase, then January 1967 is its equivalent for Cotter's generation. Three major accomplishments may be credited to this second generation and the survival and growth of an autonomous field probably depends as much, if not more, on these advances as the breakthroughs of the earlier founders.

First, this second group moved historical archaeology beyond its early nurturing but insecure and limiting housing in private foundations and governmental agencies (especially the National Park Service) into the academic world. It was only when the field gained an academic foothold during the great expansion of universities in the 1960s and early 1970s that the first historical archaeologists could not only technically but also intellectually train succeeding generations.

Second, this group went beyond the early fieldwork concentrations. Until 1960 almost all historic period excavations involved either contact-period Native American sites or *famous* ("historic") Euroamerican sites. Harrington's advocacy of the title "Historic Site Archaeology" highlights this strong but narrow focus, as does work at Jamestown, St. Augustine, Ste. Marie I, Franciscan missions, Fort Vancouver, or a Plymouth Plantation. After 1960 all time periods (not just the "earliest") and a full range of everyday rural and urban sites, as well as entire "peoples without history" (for example, African Americans, workers, women) were added to the two founding themes, thus drawing the field back into general anthropology and away from earlier preservation-restoration goals. Third, this second period cohort also firmly established historical archaeologists as a solidly organized, self-perpetuating community of researchers within general scholarship. John Cotter stood as an initiator or one of a small group of pioneering disciplinary builders for

8.2. July 1973 Summer School Class under Cotter at the Athenaeum of Philadelphia. Site of the 1775–1835 Walnut Street Prison. Archaeology students from Cheltenham High School are being shown the site by Dr. Cotter. Photo courtesy of the University of Pennsylvania Museum of Archaeology and Anthropology.

each of these three advances; and at the center of this building process was not Jamestown but Philadelphia.

In 1957 Cotter returned to the city as a NPS regional archaeologist for the Northeast and took the geographical opportunity to reactivate his earlier graduate career (1935–37) at Penn as a Ph.D. candidate in anthropology. Within two years he had retaken class work, passed his doctoral exams, and turned in a version of his Jamestown report as his dissertation. He was granted the Ph.D. in 1959. By 1960 he was an established and recognized, if newly arrived, figure in Americanist historical archaeology, and it was from the start of that decade that he used Philadelphia and its institutions (especially the NPS office, the University of Pennsylvania, and the University Museum) to solidify and expand the horizons of his newly adopted field.

In a memorandum dated June 28, 1960, he started a long personal interaction with Anthony N. B. Garvan, the chairman of the newly established Department of American Civilization at Penn, as seen in the following memo (Cotter Papers):

> The Department of Anthropology has requested my VITAE and it has been suggested by Dr. Goodenough that you may wish to see a draft of a course on Historic Sites Archaeology. Included herewith is my Bibliography, in which I have not included items in press, such as a book review for William and Mary Quarterly and an article for *Expedition,* printed by the University Museum.
> John L. Cotter

American Civilization 770 ("Problems and Methods of Historical American Archaeology"), first offered in the academic year 1960–61 and expanded three years later with the addition of a summer field component, is very likely the first class to carry the designation of "historical archaeology" in America. For just short of twenty years (1960–79), Cotter taught or helped to teach a series of lecture and field courses that explored Philadelphia as an archaeological site. He is fondly remembered by numbers of former students as a supportive and lifelong mentor (see quotes from former students Paul Huey, Gerry Wheeler Stone, and Betty Cosans-Zebooker in Schuyler 1999). He could nevertheless on occasion be a hard taskmaster (Cotter Papers):

> Letter to a Penn Student (August 18, 1967)
> Your examination and notes, although not of the first order, were con-

> siderable better than your field work. In brief, you are not disposed to become a field technician, and it is not upon what you did in the field that the grade is given. I have taken into account particularly the fact that you were in undergraduate status, and not in the same league with experienced graduate students who knew what to do to develop skill in a field enterprise. You do show an indication of analytical thinking, and can hopefully make something of archival and reference sources. If you get no more than an ability to set up a reasonable archaeological project if called upon to do so, in the future, you have earned your [grade]. But for heaven's sake, recommend a competent archaeologist to do the work. Sincerely, John L. Cotter

Across the 1960s, the instructors who taught the subject of historical archaeology were either unsecured adjunct faculty members (for example, Cotter at Penn or Arthur Woodward [1963–64] at Arizona) or standing faculty offering small, occasional graduate seminars or only incorporated historic archaeological examples into their general introductory courses. Cotter, as a visiting adjunct associate professor, reflects the first attribute; but he also, in response to Tony Garvan (American Civilization) and the nature of Philadelphia as a fully historic eighteenth- and nineteenth-century site, offered an odd course that ironically in many aspects was a full decade ahead of its time. These unusual influences at Penn pushed Cotter equally away from his initial attempts to present historical archaeology as a universal field ranging across world history from the Classical Mediterranean to Ancient Egypt to Jamestown or as a simple extension of North American prehistory (contact-sites) directly into the core of the field—European (later African and Asian) colonial cultures and their internal evolution in the New World. Historic artifact categories (ceramics, glass, metal) gained more attention than trade goods, and nineteenth-century sites eventually gained equal standing alongside colonial sites, such as Valley Forge, in Cotter's evolving class coverage. American Civilization 770, although somewhat disjointed, looked more like a current class in historical archaeology than contemporary offerings during the 1960s. Cotter was cut off from the main clusters of archaeology students at Penn. He was never appointed to the graduate group in anthropology; his classes were never cross-listed by either anthropology or classics; and even in the *Guides to Departments in Anthropology,* although "Anthropologists in Other Departments and Schools" appears as an added section in 1965, Cotter is not acknowledged by the Department of Anthropology until 1971

("John L. Cotter PhD, Pennsylvania 1959, Vstg Assoc. Adjunct Prof. American Civilization. Historical Archaeology"). Nevertheless, he made his own student converts. A few archaeology undergraduates and graduate students excavated the university catalog and found him, and he worked closely with Garvan's material culture crowd. American Civilization 770 was, for a decade, one of the most consistent and continuous offerings in American historical archaeology.

Cotter was equally important in regard to the second contribution of his generation in adding new topics to the field's subject matter, topics that helped to pull historical archaeology back into the changing core of cultural anthropology. He did not take the more obvious route and develop research that parallels the study of European-Native American contact by exploring other "peoples without history." It was Charles Fairbanks and others who added African Americans, overseas Chinese, slaves, workers, and women. Cotter was more radical. Once he had selected Philadelphia as his laboratory, he became a primary founder of urban archaeology in the United States and a secondary advocate for excavating or recording industrial sites. Building on the earlier work (starting in 1952–53) of Paul J. F. Schumacher and other NPS archaeologists at Independence National Historic Park in the heart of Old Town, Cotter moved his NPS colleagues, his Penn students, and other local researchers onto historic sites across the metropolitan region. Between 1963 and 1978, historic estates, taverns, porcelain factories, churches, forts, prisons, and urban neighborhoods were at least sampled, and gradually an outline of Philadelphia's archaeological history emerged. In 1974 he expanded even this new urban focus by borrowing the concept of industrial archaeology from Europe and assigning or urging others to add local industrial sites to the inventory. In was not until 1992 that Cotter joined with Daniel G. Roberts and Michael Parrington to produce a tour de force, *The Buried Past: An Archaeological History of Philadelphia,* the first general synthesis for any major city in North America.

The third contribution of his generation, the social high point of the decade, would seem to move us away from Philadelphia. A series of panel discussions in the late 1950s and early 1960s at American Anthropological Association and Society for American Archaeology meetings set the stage for the "International Conference on Historical Archaeology" called together by Edward Jelks, Cotter's colleague from Jamestown (with a program of papers organized by Arnold Pilling) at Southern Methodist University in Dallas, Texas. Philadelphia, nevertheless, was in the background. Cotter was one of the cofounders of the Society for Historical Archaeology (SHA) in Dallas, elected as its first president and filling in as the first editor of its new journal, and, during 1967, using his offices in Philadelphia, he organized the first official SHA Annual Meeting in Williamsburg. In January 1968 SHA was incorporated in Pennsylvania, and Cotter drove his Volkswagen Beetle carrying the first run of *Historical Archaeology* (Volume 1–1967) south to Virginia. In 1976 the national bicentennial year, he invited SHA to meet in "*one* of the largest eighteenth-century cities in the English speaking world" and again in 1982 invited them back to participate in Philadelphia's tricentennial celebration.

Philadelphia may not be the "second largest city in the English-speaking world" of the eighteenth century, and John Cotter is not (as is frequently assumed) a member of the first truly pioneer generation for the field, but the city and Cotter together during the 1960s and 1970s helped to create a holistic, institutionally well grounded and fully expanding Americanist historical archaeology directly ancestral to the discipline in the year 2000.

Acknowledgments. When John L. Cotter died on February 5, 1999, an organized effort involving Cotter, the author, and Alessandro Pezzati, reference archivist at the University of Pennsylvania Museum Archives, was under way to move his personal papers and manuscripts to the museum archives. The John Lambert Cotter Papers are now housed in those archives and have undergone a preliminary sorting and organization. Scholarship owes a great debt to Alex Pezzati for his energetic moves to save these primary sources, and I in turn completely depended on his aid in my initial research into Cotter's personal papers and manuscripts. Many of the statements in this chapter are based on these holdings. With an important exception of a solid run on his work at the Paleoindian type site at Clovis, New Mexico (1936–37), the Cotter Papers are mostly concerned with his activities in Philadelphia after 1960, especially his teaching at Penn and active field work in the city between 1960 and 1970. An equally important section carries coverage up to his passing in 1999.

9

Archaeology, Philadelphia, and Understanding Nineteenth-Century American Culture

Steven Conn

There are two stories that unfold in the pages of the essays collected here, one about a science, the other about a city. The first contributes in important ways to the developing historiography of American archaeology. This is particularly important because the writing of the history of archaeology—especially of American archaeology—has lagged behind that of many other sciences. Certainly the history of American anthropology, medicine, and chemistry, to name just three, has become more thoroughly integrated into a larger historical narrative than the history of American archaeology has. As James Snead (2001:xviii) put it recently, "the complex relationships between archaeology and society have often defied analysis."

This is a remarkable scholarly lacuna for at least two reasons. First, interest in pursuing archaeology during the nineteenth century grew almost exactly with the nation itself. Americans, at least in the first half of that century, took a nationalistic pride in the artifacts that came out of American soil. Second, that research commanded tremendous public attention. While physics and chemistry became increasingly remote for ordinary people, virtually any American who participated in the literate culture of the nineteenth century probably had the chance to read about archaeological discoveries in books, magazines, and newspapers. More than that, as local archaeological and historical societies proliferated across the country, countless more people actually dabbled in archaeology than ever set foot in a lab.

The second story presented here is about Philadelphia, about its role in the history of archaeology specifically, but about its role in the nation's scientific and intellectual life more generally. This story too has been underreported, and perhaps for some of the same reasons. Yet a more thorough understanding of nineteenth-century Philadelphia is just as important for our understanding of American culture in these years as is a more complete study of archaeology. Though these two stories are necessarily linked, I want to disentangle them, at least for a moment, to consider each on its own.

Whither American Archaeology?

To begin with, we must be more specific about the phrase "American archaeology." It has two possible meanings: archaeology done by Americans, and archaeology done in the United States, searching for answers to questions of American prehistory. As we will see, this is not merely semantic hair-splitting. Through much of the nineteenth century in the United States, those two enterprises were synonymous. Americans dug their way through the Ohio and Mississippi valleys, and eventually sites across the continent, hoping to determine exactly who had built those intriguing mounds, when they had arrived on the continent, and whether ancient Americans were related to the indigenous groups who still occupied the land. In order to place the archaeology described in these essays in its proper disciplinary context, we must reach back to the early part of the nineteenth century when excavations—or at least digging—in mounds and earthworks began in earnest.

The rush to the field during the antebellum period generated countless artifacts, dozens of publications, and at least one major institution. The American Antiquarian Society in Worcester, Massachusetts, was founded in 1812 "to discover the antiquities of our own continent; and, by providing a fixed and permanent place of deposit, to preserve such relicks [sic] of American antiquity as are portable" (Thomas 1820:18). Fittingly enough, the society's official journal was titled *Archaeologia Americana.* Without question the team of E. G. Squier and E. H. Davis, both residents of a small town in southern Ohio, produced the most important work in American archaeology during this initial period, and they enjoyed the greatest celebrity for it. Their magisterial volume *Ancient Monuments of the Mississippi Valley* appeared in 1848 and became the standard work on the topic (Squier and Davis 1848). In it, and in other related publications, the two stressed that they had conducted the work personally, thus distinguishing themselves from many armchair archaeologists who speculated on matters archaeological by reading secondhand accounts. They also outlined the rudimentary principles of stratigraphic excavation, and they insisted that archaeology ought to be empirically, rather than theoretically, driven. "Archaeological research," Squier (1849:206) wrote, "to an eminent degree, demands a close and critical attention to the facts upon which it is conducted."

Squier and Davis's volume appeared at what we might now identify as the high-water mark of this initial period of American archaeology. Squier and Davis were part of a generation of investigators writing and publishing about Indians past and present in the 1840s and '50s that included Caleb Atwater, Henry Rowe Schoolcraft, Lewis Henry Morgan, and Samuel Haven, whose book *The Archaeology of the United States* (Haven 1856) rivaled *Aboriginal Monuments* in significance. The Civil War brought this "golden age" to an end, and not simply for the obvious disruptions it caused, precisely in the geographical areas then of primary archaeological interest. Just as significantly, the Civil War stands as a formative moment in shifting ideas about American nationalism and nationhood. And changing relationships with American Indians stood at the center of those transformations.

While the postwar period in American history might aptly be called the Era of Indian Wars, it also became clear in these years—Little Bighorn notwithstanding—that the conquest of the Indians was no longer a question of whether but of when. As Indians receded as an urgent and immediate threat to most Americans, so too the field of American archaeology, which necessarily dealt with questions related to Indians, ceased to enjoy some of the public visibility it once had. Tellingly, volumes 4, 5, and 6 of *Archaeologia Americana,* for example, published in 1860 and 1874 contained much about the history of colonial America and nothing about American Indians or indeed about archaeology at all.

According to Irving Hallowell (1960:83) a "new era" in American archaeology began in the 1880s, initiated by developments at Harvard University and at the Bureau of American Ethnology. By the 1880s, then, twin torches had been passed: from institutions like the American Antiquarian and the American Philosophical Society, which had helped foster American archaeology in its incipient period, to the federal government and Harvard University; and from individuals like Squier and Davis (both dead in 1888) to people like John Wesley Powell, director of the Bureau of American Ethnology, and Frederic Ward Putnam, whose influence extended to many institutions beyond his own Harvard.

There was something else new about this "new era": "American archaeology" ceased to be synonymous with digging done on the North American continent, and by the end of the nineteenth century, American archaeology assumed a new cosmopolitanism in two important ways. First, Americans began considering how European theories of prehistory, especially the Danish and British, might be usefully applied to American finds.[1] Second, Ameri cans began sponsoring and participating in excavations back in the Old World. In the very year that Squier and Davis both died, for example, the Babylonian Exploration Fund provided support for the first American excavations in the Near East.[2]

This second development had perhaps the most profound impact on those who wanted to dig the sites of ancient Americans. In the late nineteenth century the exciting archaeological action was no longer in the New World but in the Old, no longer in the Mississippi Valley but in the Tigris-Euphrates. This work, done by English, French, German, and American excavators, captured the public's attention and commanded the serious institutional support. Walt Whitman, quoted in this volume by Curtis Hinsley, was remarkably astute—and at the same time anachronistic—to wonder in 1888 why, since "we have our schools and expeditions for Greek exploration," Americans did not "give our own evidences a chance to show themselves, too?" Plaintively, he asked, "why not open up our own past—exploit the American contribution to this important science?"

Whatever might have been the case in the antebellum period, by the turn of the twentieth century discoveries made in midwestern mounds, Florida shell middens, or even southwestern pueblo sites simply did not compete for newspaper and magazine space with the spectacular finds coming out of the ground in Egypt, the Near East, and the Mediterranean basin. North American Indian sites did not produce the monumental architecture, breathtaking arts and crafts, or ancient writing that archaeologists in the Old World seemed to be discovering every season. Even at Harvard, which among universities had the longest institutional commitment to American archaeology, the Peabody Museum found itself crowded by the collections of the as-yet-uncompleted Semitic Museum, much to Director Putnam's frustration.[3]

It was not only the spectacular nature of these finds that riveted reading Americans. These finds spoke even more directly to debates over the veracity of the Bible as a reliable historical source. In the wake of Darwinian biology and German "higher criticism," Christian religious belief found itself under dramatic assault in the late nineteenth century. Into this crisis of faith stepped the archaeologists, many of whom believed that by discovering the physical remains of Old Testament stories they would undergird the Bible as a reliable historical document and help save Christianity from the corrosive effects of scientific doubt. As Bruce Kuklick (1996:24) has put it: "By the 1880s, scholars of the Near Orient in the United States saw their studies in the front lines of a defense of the Old Testament."

There is, of course, a deep irony in all this, even if it went largely unnoticed: after all, that first generation of American archaeologists had worked mightily to understand American Indians, ancient and modern, within the historical frameworks provided by the Bible and other classical texts. Whatever the intellectual square peg/round hole dilemma of their work, theirs was an attempt to marry the empirical evidence coming out of the ground with a biblical view of the world. By the end of the nineteenth century, it was American expeditions to the Near East in defense of the Old Testament that took on this urgent, almost crusading excitement. As a result, those who sought to reconcile science and religion no longer took much interest in the archaeology of ancient America.

Likewise, although archaeologists of ancient America had to fight hard battles to create space for themselves in the newly emerging research universities, several universities created whole departments of Oriental studies or Near Eastern studies to house archaeologists and other scholars of the ancient Near East and Mediterranean. In addition, such researchers might find an institutional home in departments of classics, art history, or even history. Conversely, by the end of the nineteenth century, American archaeology had become a subset of American anthropology; as a result, those archaeologists found jobs almost exclusively in departments of anthropology and thus shared an institutional home with people interested in Indonesian folklore and African kinship patterns.[4]

Depending on how one chooses to look at this, the institutional trajectory of American archaeology reflects either the growing cosmopolitanism of American scholars and universities, which now had the resources and the confidence to participate in European intellectual projects, or the failure, despite initial hopes, to create a new and lively field of intellectual pursuit that centered around the study of Native America and of which European scholars would be jealous. One way or the other, although studies of American Indians in general, and of American Indian archaeology in particular, were among the oldest of the nation's intellectual pursuits, they did not become institutionalized in the new American research university the way that other disciplines did.

Rewriting Archaeology's Forgotten History

This, then, is the context in which we need to consider the careers of some of the actors who traverse the pages of these essays. Charles C. Abbott, whose sad career at the University Museum is detailed in this volume by David Meltzer, was one of those who carried on the legacy of the very first American archaeologists. Sifting his way through those Trenton gravels, Abbott was really trying to answer that most basic and still unanswered question posed by the presence of Native Americans: how long ago did the first Americans arrive on the continent? His inability to find what he regarded as adequate institutional support for his researches, either at Harvard or at Penn, is not merely testament to what surely seems like an irascible and grating personality (after all, when have these qualities posed an obstacle to university advancement?). More than this, Abbott stands as one of those from an older generation of what we would now call "amateur" scientists unable to fit into the new, professionalized world of the research university as it emerged in the late nineteenth century.

Clarence Bloomfield Moore too, as Lawrence Aten and Jerald Milanach have portrayed him, looks to us now like an archetypical gentleman-scholar. Complete with family money to fund trips to Europe, Central and South America, and up the Amazon, in addition to his extensive excavations in the inhospitable wilderness that once was Florida, Moore played this role as if sent straight from central casting as he navigated his customized paddle-wheel steamboat through the Florida swamps in search of archaeological sites.

The careers of Sara Yorke Stevenson and Lucy Wilson, as detailed in this volume by Elin Danien, Eleanor King, and Frances Joan Mathien, remind us that one hundred years ago archaeology offered women greater opportunities than did many other scientific pursuits. At one level, Stevenson and Wilson stand on either side of the American/Old World divide I discussed earlier; Stevenson's scholarly interests were largely Egyptian while Wilson worked in the Southwest. At another level, both their careers illustrate the still precarious institutional position women occupied at the turn of the last century. Stevenson was instrumental in helping build Penn's museum, but in large measure this came through her extraordinary partnership with Provost William Pepper. Her work with the museum did not survive his death by too many years. She soon found herself resigning from its board and accepting a curatorship at the Pennsylvania Museum, now the Philadelphia Museum of Art.

Wilson's work in the Southwest found a public audience through an exhibit at the Philadelphia Commercial Museum in 1918. While the Commercial Museum was once an enormous and influential operation, it was hardly a center of archaeological research or display. Further, by 1918 it had already begun its long, slow slide from a position of importance to one of obscurity.[5]

None of these figures makes it into our conventional narrative of the history of archaeology, such as it is, and Daniel Brinton's work as an anthropologist, which Regna Darnell has done so much to recover, may best illustrate why that is. Hinsley is probably not alone among historians of archaeology and anthropology who remain "mystified" by Brinton's career. For most historians of anthropology, Brinton exists as a quirky footnote: yes, he did technically occupy the first university professorship of anthropology; but no, in fact, this did not have any discernible impact on the future of either the university or the discipline.

Brinton was not a part of the professionalization and institutionalization of anthropology and archaeology taking place before his eyes. He built no department, nor did he train any students; he accumulated none of the capital that would have currency in the new world of university-based science. In fact, he probably had a greater public influence through the lecture courses he offered at the Academy of Natural Sciences. In addition, insofar as Brinton can be said to have had a research "agenda," it was American. Despite his belief that "American archaeology will in time rank equal with that of Egypt and the Orient," at Penn he found himself on the wrong side of the debate over what that museum ought to be.[6]

In these senses, Brinton stood on the far side of the generational divide I mentioned earlier. A polymath, by the time he took his "professorship" at Penn he was quite literally the senior figure in American anthropology, and his career can be said to be more deeply rooted in nineteenth-century intellectual traditions than to have reached toward those of the twentieth. Brinton, as an amateur and self-taught anthropologist committed to studying American questions, who was more at home in the lecture hall of the academy than in the seminar room of the university, had become something of an anachronism by the end of his life. For those looking for the ancestry of modern anthropology, therefore, Brinton is a distant relative.

This, I think, helps explain why these researchers, and others, have by and large been written out of the history of American anthropology and archaeology. As I mentioned at the outset, the history of American archaeology and anthropology (in the United States in particular the two are inextricably linked), as opposed to the history of other sciences, or even of European archaeology, remains an underdeveloped field. That which has been written tends to the teleological, which is to say that it takes the current state of American archaeology as an inevitable given and looks to the past to establish a genealogy that brings us up to that present. As George Stocking (1987:xiii) writes in the introduction to his book *Victorian Anthropology,* the historiography of anthropology tends "to trace ideas backward in time in order to establish lineages—albeit sometimes interrupted—for contemporary theoretical viewpoints." Perhaps Curtis Hinsley (1989:80) put it better when he wrote that the history of archaeology "still remains essentially Whig history."

As a case in point, when Gordon Willey (1973) wrote the introduction for the republication of Samuel Haven's (1856) *Archaeology of the United States,* he described its importance as marking "the divide between archaeology as romantic speculation and the beginnings of disciplined description and classification." The historiography of archaeology can, without too much messiness, be divided into three kinds. The first traces the broad history of archaeological pursuit—from the renewed interest in the classical world during the Renaissance, through Enlightenment antiquarianism, to the revolutions in geology during the nineteenth century, arriving finally at a scientifically based archaeology. Brian Fagan (2000) sketched such a history in his *In the Beginning: An Introduction to Archaeology,* and Glyn Daniel, perhaps the most prolific historian of European archaeology, has written several versions of this narrative in his many books (for example, Daniel 1976).

The second variety of history might be called regional histories. Studies of this type constitute the bulk of writing on archaeological history, and they tend to trace the history of archaeological research in particular areas as a way of illuminating how our current understanding of those areas was reached. While such histories written about the Tigris-Euphrates, Egypt, or Greece trace work done back into the eighteenth century, archaeological work has now spread all over the globe and hence there is virtually no corner of the planet that does not now have its own regional archaeological history.

Finally, there is an important literature that tries to sort out the often complicated relationship between history, science, and archaeology. English archaeologists in particular have seemed inclined to see archaeology as a way of navigating between humanistic history and hard-core science; many

American archaeologists, especially in the wake of the "New Archaeology" of the 1960s, have sided squarely with science. "Archaeology," according to D. B. Banforth and A. C. Spaulding (1982:193), has "the essential character of science, not history."

Each of these kinds of history is driven by a sense of how things are now—how we developed the careful, stratigraphic methods we now all use in our excavations; how we arrived at our current understanding of the archaeological sequences in particular places; how American archaeologists quit the business of "romantic speculation" and finally got "disciplined."

That so much of the historiography of archaeology has this teleological quality should come as no real surprise. While few who write the history of medicine are practicing physicians, or those who write the history of chemistry laboratory scientists, the writing of archaeology's history has been done primarily by archaeologists—insiders writing from an insider's point of view.

Writing the history of how things have come to be as they are, of course, is a perfectly legitimate, important project, and there are considerable and obvious advantages to a history written by insiders: practitioners can be more attentive to issues of technique, significance, and even personalities than outsiders. At the same time, however, such a history has limitations. Sometimes the historical questions that get asked remain internal to the current state of the profession—hence the teleological drift, the how-did-we-get-to-where-we-are-now concern with so much of the history of archaeology.[7] Related to this in the search for genealogy has been the impulse to distinguish what past researchers got "right" and what they got "wrong." Exemplary of this kind of history is Stephen Williams's (1991) book *Fantastic Archaeology,* subtitled *The Wild Side of American Prehistory.* In it Williams takes the reader on a delightful romp through what he calls the "fringe" of serious archaeological work. But his implication is that much of what went on in the archaeological fields during the nineteenth century cannot be taken seriously by scholars of American prehistory today.

Quite right, but incompletely so. While much of the history of archaeology may no longer be relevant to archaeology today, and is thus treated perfunctorily or even derisively by those who write that history, it may prove to be quite important to help us examine a host of other kinds of historical questions. An expanded history of archaeology would place it in the larger context of American cultural and intellectual history. The history of archaeology might illuminate the relationship between science and nationalism in the nineteenth century; it might shed light on how perceptions of and responses to Native Americans changed during the period. Late-nineteenth-century archaeology in the Near East, as Kuklick has demonstrated, was intimately connected to theological debates going on in those years; the same is true, I suspect, of American archaeology earlier in the nineteenth century. Rather than worry about the distinctions between "professionals" and "amateurs," we might use a more broadly conceived history of archaeology as a marvelous opportunity to examine how those distinctions got made in the first place and the implications of those distinctions for the way intellectual discourse developed. The careers of Sara Stevenson and Lucy Wilson underscore that women played an important role in archaeology at the turn of the last century and that looking at the intersection of archaeology and women's history might be very interesting indeed.

In short, the history of archaeology needs to include those who got it wrong as well as those who got it right, the evolutionary dead ends as well as the close relatives. Defining the history of archaeology narrowly to include only the ancestors of today's researchers misses the full significance of archaeology for Americans in the nineteenth century. Such a history will have to make room not only for Samuel Haven and Frederic Putnam, but for Charles Abbott, Henry Mercer, Lucy Wilson, and Clarence Moore as well. These essays take some important first steps in that direction.

Philadelphia Makes the Modern

Shortly before I sat down to write this essay, I finished Andrea Barrett's (1998) novel *The Voyage of the Narwhal.* Set in the mid–nineteenth century, the book tells a gripping story about Arctic exploration and how it drove men to behave. It's a fun read, but it isn't my purpose here to make bedtime reading suggestions. Rather it is to point out that this historical novel, spun around several real events, is set in Philadelphia; the Academy of Natural Sciences, for example, is a central part of the stage set upon which the characters play out their drama.

The Voyage of the Narwhal stands as a small reminder that through the nineteenth century Philadelphia was still at the center of the nation's scientific endeavors. This will come as a surprise to many who believe that once Benjamin Franklin died, nothing much of scientific or intellectual consequence came out of Philadelphia. Indeed, the way the cultural and intellectual history of the United States has been written shares some of the same teleological tendencies as the historiography of archaeology. For too many writers, the story must end in New York, and so only those cultural and intellectual roads that lead there are worth traveling.

Robert Schuyler begins his essay here by poking fun at the widely held belief that by the late eighteenth century Philadelphia had grown to be the second largest English-speaking city in the world (albeit a distant second to London). Whether or not that tidbit is factually true, or even provable one way or the other, it has contributed to a larger perception that Philadelphia's moment came at the Revolution and went once the national capital moved to Washington in 1800. Without in any way downplaying the significance of the city's role in founding the nation, I would suggest that Philadelphia's nineteenth-century history is at least as interesting and important, and that it has been remarkably ignored.

In a host of ways, from its boundaries to the patchwork of its neighborhoods to its constellation of institutions, Philadelphia is a creation of the nineteenth century, and more specifically of the industrial economy that blossomed there between the Civil War and World War II. That economy, which grew to encompass perhaps the greatest industrial diversity of any American city, is itself worth studying from the perspective of the history of science and technology. Philadelphia in the nineteenth and early twentieth centuries was a leader in what we would today call applied science—the use of technological and scientific innovations on the factory floor. There are several studies of Philadelphia's working people and of the city's economy during these years. But only one of which I am aware, Bruce Sinclair's (1974) book on the Franklin Institute, *Philadelphia's Philosopher Mechanics,* examines the relationship between mechanic's institutes and the growth of Philadelphia's industries—and his book only takes the story to the Civil War.

Beyond this, however, Philadelphia's intellectual life was just as lively in the nineteenth century. That life revolved first around institutions like the American Philosophical Society, the Library Company, and the Academy of Natural Sciences; later those centers were joined by the University of Pennsylvania, the Commercial Museum, and other places besides. While Philadelphia now shared the intellectual spotlight with Boston, New York, and Washington, it by no means ceded it altogether as the nineteenth century wore on. Indeed, in my book, *Museums and American Intellectual Life, 1876–1926* (Conn 1998), I argued that Philadelphia was on the cutting edge of museum building during these years and built a collection of institutions without equal anywhere else in the country.

Penn's museum, through which all the figures in these essays surely passed in one way or another, exemplifies the way in which Philadelphia's museums were not simply built as repositories for the vast accumulations made by acquisitive collectors. Rather, Penn tried to create institutional space for the new and emerging fields of anthropology and archaeology (and it thus reinforced the deep connection between the two). In most other cities, notably New York, Chicago, and Washington, such collections were never liberated from their institutional homes in natural history museums. Penn's museum, along with the Philadelphia Museum of Art (originally the Pennsylvania Museum), the Commercial Museum, Henry Mercer's museum, and the venerable museum at the academy, stood as volumes in a veritable encyclopedia set of museums. That set attempted to give institutional form to different bodies of knowledge—natural science, American history, art, anthropology, economics—and to put that knowledge on display for the public.

In the world of arts and letters, Philadelphia is seen as entering a long, sleepy twilight after the Civil War. The state of its cultural life merely reflected a larger sense of the city itself—complacent and staid according to some observers, content and corrupt according, famously, to one. By the turn of the twentieth century, New York had become the center of "modern" life, the heart of the nation's cultural and intellectual life. This is so obviously true that it hardly needs recounting. And yet here too I think this truth is incomplete. As some scholars are now beginning to explore, whatever constitutes the phenomenon of American "modernism" must really be understood as various and polyphonous. The narrative of modernism that leads to New York, therefore, however significant it may be, is only one of several stories to be told. That American modernism and New York are seen almost synonymously reflects the slant and predispositions of certain critics and historians as much as it reflects the ebb and flow of cultural currents. As much as the Gotham narrative includes, it ignores the realist writers and painters of Chicago; scientific developments in a host of institutions; and even Frank Lloyd Wright, whose antiurban sentiments have made him hard to place in the context of New York modernism.

So too, I believe, there is a Philadelphia modern, different to be sure from that of New York, that has gone incompletely reported. Elizabeth Johns (1983) has demonstrated powerfully how Philadelphia painter Thomas Eakins linked his art to the "heroism of modern life" in Philadelphia. Whether in his two magnificent medical portraits, *The Gross Clinic* and *The Agnew Clinic,* or in his scenes of rowers on the Schyulkill, Eakins's works are records of Philadelphia's role in shaping the modern world. The Pennsylvania

Academy, where Eakins taught, also produced John Sloan, the foremost member of New York's "Ashcan School," which came to prominence in the early twentieth century, and which constituted briefly the American avant-garde. Charles Sheeler, for my money the most interesting American modernist painter, also trained at the academy and, as I have argued elsewhere (Conn 1998), his particular modernist vision was shaped by a formative friendship with none other than Henry Mercer.

Thomas, Lewis, and Cohen (1996) have forced us to reconsider the development of twentieth-century American architecture by tracing its line back through Will Price to Frank Furness, perhaps the most inventive architect of "modernity" America produced. Any history of American architecture that relegates Philadelphia to provincial and conservative status, as most do, fails to explain how the PSFS (Philadelphia Saving Fund Society) Building, considered by many the first international modern-style building to go up in the United States and arguably still among the finest, got built in 1932 at Twelfth and Market, not in Chicago or Manhattan (Kostoff 1995:716, Conn 1998:182–86). Nor can it explain the line of architects that grew through the twentieth century to include Louis Kahn, and the "Philadelphia School" that grew up around him, and more recently Robert Venturi. Indeed, it can be argued that, just as the international modern style made its debut in Philadelphia, postmodernism was born in Philadelphia when Venturi (1966; Venturi, Brown, and Izenour 1977) published his vastly influential books, *Complexity and Contradiction in Architecture* and *Learning from Las Vegas.*

I don't mean to be making a parochial, booster's argument here, though it may sound a bit like that. My point is to suggest that much of the story of Philadelphia's intellectual and cultural life in the post-Franklin, post-Rittenhouse era remains to be told. A quick bibliographic survey makes my point: while there are at least half a dozen biographies of Harvard's Louis Agassiz, Darwin's chief antagonist in America, there is only one that I can find of Philadelphia's Joseph Leidy, and that book is quite recent. Leidy is a fascinating and deeply important figure in the history of natural science and of medicine, splitting his efforts between Penn and the academy. He was, as the wonderful subtitle of this biography calls him, "the last man who knew everything" (Leonard 1998). Likewise, figures as diverse as C. S. Rafinesque and Edward Drinker Cope have yet to receive the full scholarly consideration that their careers deserve. While I count at least three scholarly biographies each of the University of Chicago's William Rainey Harper and of Johns Hopkins president Daniel Coit Gilman, I can find only one—and quite old at that—of Penn's William Pepper, an educational leader and institution builder of such energy that he was referred to in his own lifetime as the second Benjamin Franklin.

There are doubtless several reasons for this historiographical neglect. But Schuyler may have hinted at one of them in his essay on the career of John Cotter. Cotter arrived in Philadelphia to work for the National Park Service in 1957, less than ten years after work had begun to create Independence National Park. While the creation of the park is itself a fascinating story of destruction, invention, and mythologizing, as well as of archaeological recovery, it has been a central part of shaping Philadelphia's late-twentieth-century image as that of an eighteenth-century town.

There is a delicious irony in this: just as the city's industrial economy—the economy built in the nineteenth century—was beginning its inexorable deterioration, the city staked its postindustrial future on attracting tourists to the eighteenth-century shrines preserved now in Independence National Park. Say "Philadelphia" to any ordinary American, and "Independence Hall" is usually the Rorschach reference that pops out. No accident that the city officially measures its tourist volume by counting people who visit the Liberty Bell. It may be, then, that historians over the past half century have internalized this popular view of Philadelphia as a place where one goes to study the eighteenth century but not much else beyond that.

Attempting to recover Philadelphia's nineteenth century intellectual and cultural history might be a way to explore several ideas. As America's intellectual life became professionalized and institutionalized at the turn of the last century, it was driven by a cosmopolitan goal. The creation of knowledge transcended the particularities of place, and Americans participated in creating an intellectual discourse that was not merely national but trans-Atlantic as well. Philadelphia shaped that discourse in ways we don't yet fully know, and that story needs to be told.

And yet at the same time, my own sense is that locality continued to matter in America's intellectual life more than the cosmopolites wanted to acknowledge. While historians, anthropologists, physicists, and a host of others organized themselves in professional associations designed to create disciplinary communities outside the boundaries of colleges and universities, professionalization and institutionalization went hand in hand in the late nineteenth century. They certainly shaped each other, and Philadelphia's long institutional history was bound to have given a particular shape to the way newly trained professionals operated there. William Rainey Harper might have been able to start tabula rasa at the University of Chicago without any institutional history upon which to build, or any baggage to weigh him down, but the American Philosophical Society was already 150 years old by the time the University of Chicago was founded in 1893. Finally, restoring Philadelphia's nineteenth-century intellectual and cultural history might prove important for the city itself. Philadelphia is a great deal more than a few square blocks of eighteenth-century buildings. Philadelphians themselves are largely unaware that there was life after Franklin; the rest of the nation is too. Whatever Philadelphia's future may prove to be, it must grow out of a more complete sense of its past.

I began this essay by saying that this collection told two intertwined stories—one about archaeology, the other about a city. Perhaps it would have been more accurate to say that the essays in this volume suggest two interrelated areas of research: one, a more complete, more contextualized history of American archaeology; the other a rediscovery of the role Philadelphia played in shaping the nation's nineteenth- and early-twentieth-century intellectual discourse. Each powerfully shaped the other, and our understanding of America's intellectual life will remain incomplete without both.

Notes

1. The term "prehistory" was coined by Scottish archaeologist Daniel Wilson (1851). It and he came across the Atlantic shortly thereafter, and the words "prehistory/prehistoric" were in common use in the United States by the 1860s and '70s (Kehoe 1998), though I have been unable to locate the exact first usage of the word in this country.

2. For a thorough account of this, see Kuklick (1996:esp. chapter 1). While there had been a few Americans to work in the Old World during the antebellum period, it is certainly true that this activity increased dramatically during the last quarter of the nineteenth century.

3. See Putnam to Charles Eliot, December 21, 1889, Peabody Museum Archives, Director's Records, Box 5.

4. My research on this topic is largely anecdotal and by no means thorough, but my sense is that even today, art history departments don't generally reserve faculty slots for experts in Native American art except at a handful of places. Likewise, though language study in this country began with the study

of Indian languages, there are no departments in American universities devoted to teaching them, as there are with, say, Slavic, Romance, and classical languages. See Andreson (1990:esp. 176–88).

5. For more on the Commercial Museum's history see Conn (1998:chapter 4).

6. Brinton to Jeffries Wyman, quoted in Hinsley (1989:84). For more on the fights at the University of Pennsylvania Museum see Conn (1998:chapter 3).

7. There are, I hasten to add, some obvious and important exceptions, including the work of Curtis Hinsley, James Snead, Bruce Trigger, and Alice Beck Kehoe.

10

Philarivalium

Alice Beck Kehoe

We have here an ethnography of a segment of Philadelphia society between the Civil and World Wars. Illuminated by the burning souls of restless nonconformists, Philadelphia appears a city of intense rivalries fed by tidal currents of economic change. ("Rival," from Latin *rivalis,* derived from *rivus:* one who draws from the same river as another.) Some, such as Abbott and the blacks hired by Clarence Moore, would be hewers of wood; others, notably Daniel Brinton, would sit in armchairs by fires fed by that wood. Women of the moneyed class, notably Sara Stevenson, but including Lucy Wilson, maneuvered with and through men to embody their visions. Chapters here vivify contrasting eras, that of the bulk of this book when archaeology was shifting from gentlemen's avocation to professional status, and three generations later when discipline boundaries opened to accommodate the cultural diversity reflecting a vastly more democratic America.

Philadelphia is peculiarly significant for an ethnography of a critical shift in the practice of American archaeology. Back in July 1776, Thomas Jefferson, Benjamin Franklin, and other well-informed scientific observers had gathered in congress there to declare,

> When in the Course of human events, it becomes necessary for one people to dissolve the political bands which have connected them with another . . . a decent respect to the opinions of mankind required that they should declare the causes which impel them to the separation. . . . The history of the present King of Great Britain is a history of repeated injuries and usurpations . . . To prove this, let Facts be submitted to a candid world. . . .

> He has excited domestic insurrections amongst us, and has endeavoured to bring on the inhabitants of our frontiers, the merciless Indian Savages, whose known rule of warfare, in an undistinguished destruction, of all ages, sexes and conditions [27th in the list of 27 "Facts"].

Who would dispute the fact established by such hallowed authorities, that the indigenous inhabitants of our continent were merciless savages?

A century after the Continental Congress, Philadelphians struggled with the questions outlined by Fowler and Wilcox in their introduction herein. Charles Abbott doggedly insisted on relating the Trenton gravels to "the Drift" identified by pioneer European glacial geologists and artifacts in the gravels to an American Paleolithic. Abbott exemplifies the nineteenth-century naturalist who recognized specialists but did not hesitate to collect, classify, and interpret across the spectrum of nature. Abbott's vicissitudes illustrate the pragmatic difficulties of the early Smithsonian's Enlightenment ideal, voiced by Joseph Henry, of a "democratic science" (Hinsley 1981:151–53). A lovely ideal, every citizen—Jefferson's yeoman farmer—contributing to scientific knowledge by his careful, systematic field observations, conveyed to a central repository in the nation's capital. A conflicted reality, especially when the yeoman has not slaves to work his farm. The construction of disciplinary boundaries and professional career paths was a means of democratizing science, responding to the need to reduce conflicts between obtaining a living and pursuing scientific research. Stubborn mavericks like Charles Abbott will always be with us. The twentieth century opened opportunities to more reasonable men, and finally women, to work in science simply as educated professionals, neither driven like Abbott nor leisured like Moore.

Sociology of professionals emphasizes autonomy to be a major value prized by the professional class (Kehoe 1999:4). This runs as a leitmotif through these chapters. Autonomy may be more a goal than achievement, given credentialing institutions' authority plus agendas and limitations expected by funding grantors. Stephen Dyson remarked of classical archaeologists, "Those who had a sugar daddy dug, and those who didn't, regardless of the worth of the project and the creativity of the archaeologists involved, didn't" (Dyson 1999:112). Clarence Moore was his own sugar daddy. Abbott's rival Henry Mercer was financially independent, yet his experiences with the University of Pennsylvania and its museum show how chimerical autonomy is, even for men with independent incomes: he abandoned disappointing excavations in the Delaware valley and museum intrigues to fabricate unique (therefore autonomous) expertise in technical experimentation, historic artifacts, and, eventually, leadership in the Arts and Crafts movement. Moore on the *Gopher* and its predecessors looks the fantasy of autonomy, steaming where and when he pleased; even so, and ignoring thc inexorable demand for boiler wood, Moore could not break out of the bounds of his era's conventional expectations. Aten and Milanich (this volume) insightfully note that Moore "constrained, by the extent of his digging, what can ever come to be known about [his] subject." The originating social, temporal, and geographic contexts of a discipline, happenstances of who is drawn into it, their luck or mishaps, and ideological and economic developments in the larger society, all affect the vaunted autonomy of scientific research.

Sara Yorke Stevenson epitomizes the envied status of a rich, socially prominent, admired lady. So brilliant as a child that she insisted her parents enroll her in a more intellectual school, Sara Stevenson benefited from a cosmopolitan upbringing fitting her for the salons of the intelligentsia. A study of wealthy women volunteer leaders in San Francisco by Arlene Daniels (1988) offers strong parallels to Stevenson's, and to Lucy Wilson's, careers. Daniels particularly stresses that the seventy upper-class women in her study have careers, albeit unpaid and, in her word, "invisible" (not publicly acknowledged to be careers). The women see their work as altruistic efforts toward community welfare, extensions of the self-effacing, nurturing behavior valued in women. Daniels sees the women as entrepreneurs and executives in a profession unrecognized because unrecompensed.

> The ease with which women of the privileged classes can make connections to the city political and business leaders, the resources they may use to advance volunteer projects, the opportunities they have to become renowned in their volunteer work comprise a particular set of assets related to their class position [Daniels 1988:269].

Daniels singles out the créme de la créme of this class, women,

> known for their speaking ability, intuition and good judgment, patience with information-seekers, knowledge of the community, and all-around acumen. . . . These are women with that extra quality that makes them stand out when they chair a meeting, manage a complicated merger, outface an opposition group, charm or neutralize an antagonist in a public confrontation [Daniels 1988:91].

There is, we see, a template for Sara Stevenson. There is also a template for Lucy Wilson, a woman below Stevenson in intellectual brilliance and charisma, in wealth, in social position, and—related to her lesser wealth and societal status—with the stigma of an overt career.

Daniel Brinton competed with Stevenson in the invisible careers business. Hinsley quotes him, "I deliberately left a profitable business that I might, on a modest competence, pursue my life as an observer, a thinker and an unpaid writer." Not, of course, in the womanly role of self-effacing nurturer, Brinton's rebuff to Stevenson encapsulates the contrasting personae appropriate, respectively, to men and to women espousing occultated careers. (Darnell calls him, herein, an "avocational anthropologist.") Brinton was a savant drawing his impressive erudition from pristine texts collected by others. His conviction that there is a general human nature, to be discovered within a planetary compass, is very much an Enlightenment position, and his armchair research very much in the mode of European philosophes. This made him an anomaly in America, where democratic science carried out through field work, by Squier and Davis, Lewis Henry Morgan, John Wesley Powell, and many others, proffered the man of science as a man of action (and yes, a *man* [Kehoe 1999:117–18]).[1]

It is Frank Hamilton Cushing, of all the principals in this book, who most emphatically embodied (literally) the active empirical mode of scientific research. Physically immersing himself in the field, Cushing utilized all his senses, tactile, taste, and kinesic as well as visual and aural. He and others speak of his intuition—better to term it subliminal perception, the challenge of articulating to others suites of sensory impressions not necessarily conforming to English language image schemas. Cushing transcended the boundaries set between humanities and science; indeed, he highlights the arbitrariness of these boundaries. His urging ethnographic experience upon young archaeologists in training is at once obvious and radical, obvious to anyone who can appreciate the value of the full panoply of sensory and intellectual experiences, radical because it admitted non-Western technology and cultural formulations to the position of preceptor rather than object.

"The inhabitants of our frontiers, the merciless Indian Savages," must be conquered and ruled. Cushing went to Zuñi *only three years after Custer's defeat,* right at the cusp of the final campaigns to affirm, "with a firm reliance on the protection of divine Providence" (as they said in Philadelphia in 1776), America's Manifest Destiny. Roger Kennedy remarks that the moral discomfort engendered by the United States' conquests of First Nations "was lightened by denying any redeeming virtues—such as architectural skill—to the enemy" (Kennedy 1994:237). Cushing's willingness to defer to Zuñi beliefs and praise their way of life, cultural relativism before Boas, ran counter to the prevailing attitude. Both his *premise,* that a First Nation could have developed a fully valid culture deserving respect, and his *method,* mastery of the skills of that culture through participation, were more than unorthodox, they were antiorthodox. Frederick Webb Hodge was a conventional American, good at administration rather than knowledge production. The conventional archaeology he favored followed the "Baconian method" popular in American science (on Baconianism, see Barnes, Bloor, and Henry 1996:143–47; on its popularity in America, see Bozeman 1977 and Kehoe 1995). Data are to be collected and classified into "natural histories," free from the distortions imposed by a priori hypotheses. A scientific archaeology in America would dig up artifacts and map ruins, classify them according to similarities and differences, and publish the results as a contribution to the natural history of a region, along with its geology, zoology, and botany. Baconian science accommodated Manifest Destiny ideology, relegating American Indians to the status of fauna. It disallowed Cushing's cultural relativism.

The battles in Philadelphia, Boston, Chicago, and other cities over allocating museum funds to American Indian research went beyond the issue of aesthetics, and beyond the debate on whether the public should be elevated by exposure to the pure harmonies of chaste classical art, or might be debased by seeing pagan crudeness. The fundamental, dangerous issue was the contradiction between the noble universal principles declared to be America's reason for being and the reality of denying their birthright to slaves, American Indians, Asian immigrants, and the poor. Spencerian cultural evolutionism legitimated the Anglo classes' harsh domination, as Henry Adams recalled. "Unbroken Evolution under uniform conditions pleased every one. . . . Such a working system for the universe suited a young man who had just helped waste five or ten thousand million dollars and a million lives, more or less, to enforce unity and uniformity on people who objected to it; the idea was only too seductive in its perfection" (Adams 1918 [1907]:224–26). When Frank Cushing restored the beauty of the shell toad, he revealed a redeeming virtue in the forebears of the subjugated, disenfranchised Indians. America could not tolerate such subversion; Baconian science was bound to the "Facts" declared by the Founding Fathers, gentlemen whose words could not be impugned (Barnes, Bloor, and Henry 1996:146).

Schuyler's account of John Cotter's development of urban archaeology has the larger significance of indicating the tie between the democratization of higher education in the United States, primarily as a result of the G. I. Bill (Kehoe 1998:136, 212; Ryan and Sackrey 1984), and the broadening of archaeological research to include sites representative of hitherto disregarded classes. Schuyler considers Cotter's attention to a city as a whole, rather than on a class such as African-American or overseas Chinese, to be more radical than the innovations of scholars focusing on such designated classes. This is an argument to ponder, for it challenges the positivist assumption that named categories are real and bounded (for a discussion by a contemporary historical archaeologist, see Spencer-Wood 1999). In a sense, Cotter's approach is cultural ecology to the max, and in that sense part of the post–World War II scientific fashion in America (Kehoe 1998:109), but because "cultural ecology" became a label for a reductionist use of systems theory, chest-beating theorists overlooked the sophisticated grasp of Cotter's work.

Cotter's marginal position to recognized anthropology, seen in Schuyler's finding that neither he nor his courses were listed within that discipline at Penn, may be further example of the divorce between American archaeology and cultural anthropology perceived by Pinsky (1992). She contends that after Powell's death, the conservative leaders of the Bureau of American Ethnology endeavored to stem the advance of cultural relativism campaigned for by Boas. They, like many conservatives, feared a flood if immigrants would overwhelm American culture, and Boas, an immigrant, made an obvious target. Their extraordinary zeal, no doubt born of insecurity, leaps from this letter discovered by Pinsky. "A new chairman of the [National Research] Council must be selected and it is most important that he should not be of the Hebrew kind . . . [Clark Wissler] is . . . a two-faced Jew of the most cunning variety and an understudy of Boas" (Letter from W. H. Holmes to C. Walcott, 1921, Secretary's Correspondence, Box 62 Folder 1-2, National Anthropological Archives; I thank Valerie Pinsky for sharing this letter with me).

Wissler was a seventh-generation English Protestant American! Although Boas's influence on cultural anthropology, established through placing his students in the slowly expanding university positions in anthropology, could not be gainsaid, it has little effect on archaeology. Archaeologists could go on collecting and classifying in the natural-history mode, their distinctive orientation abetted by Boas's privileging indigenous-language texts—not to be found archaeologically in Anglo-America—over material artifacts (Darnell 2001:327). Out in left field in Boas's ballpark, archaeologists had their own game going.

Until the 1990 Native American Graves Protection and Repatriation Act (NAGPRA) gave the ball to a hitherto-sidelined team from First Nations, mainstream American archaeologists operated in what they believed to be a predominantly scientific mode. Geology, the sister excavating discipline, was their model, and the natural sciences as a group their referents. Mavericks concerned with wider cultural questions, whether loud like Walter Taylor (1948) or quiet like John Cotter, could not broach the conservatives' fort at the frontier inhabited, the Founding Father told us, by Indian savages. Chief Justice John Marshall has reiterated, in *Johnson v. McIntosh* (1823) that "the tribes of Indians inhabiting this country were fierce savages, whose occupations was war, and whose subsistence was drawn chiefly from the forest. To leave them in possession of their country, was to leave the country a wilderness; to govern them as a distinct people, was impossible" (quoted in Williams 1990:323, n. 133). This chartering myth for the United States has been the paradigm for American archaeology.

The present volume is of great value in exhibiting how contested the paradigm has been, even in the hallowed home of Independence Hall. The very Baconian, democratic science that obscured the conservative political ideology underlying American archaeology provided the empirical data confounding the myth. Careers foundered on intractable congeries of observations, not only Abbott's refusal to reconsider the Trenton gravels, but those of his rival Henry Mercer, and Cushing and Stevenson and Cotter when they stepped outside the normal science of the dominant paradigm. Juxtaposing the highly contested rival agendas in the formative years of professional archaeology in Philadelphia opens out the terrible confrontation between the founding fathers' charter myth and the First Nations they brooked no interference in dispossessing. American archaeology cannot avoid political reverberations, whether on the local scale as in Philadelphia a century ago, or today.

Note

1. The model derives from a parallel mode associated with the scientific revolution (Shapin 1996:90). Shapin's description of "the scientific author" shows a consensus on the authorial figure, whether philosophe or "mechanick": he "appeared as disinterested and modest, not concerned for fame and not affiliated with any school of grand philosophical theorizing" (Shapin 1996:108). Brinton clearly saw himself as such a figure.

Appendix

Turquoise Encrusted Toads and Raptorial Birds in the North American Southwest

Data assembled by Jenny Billideau (1986) and David R. Wilcox

Flagstaff area: A.D. 1150–1300

No.	Site Name	Bird/ Toad	Back	Perf	Comments
1	Ridge Ruin	bird	lac	no	McGregor 1943; Magician burial; near left elbow
2	Ridge Ruin	bird	shell	NA	Magician burial; bird on shell bracelet above head on right
3	Elden Pueblo	toad	shell	?	with child burial (stripe down back)
4	Flagstaff area	toad	shell	yes	Babbitt collection; with burial; has possible anus
5	Flagstaff area	circle	?	?	Babbitt collection
6	Flagstaff area	bird	wood?	?	along Leupp Road in rock piles
7	Flagstaff area	bird	?	?	along Leupp Road in rock piles

Prescott area: A.D. 1150–1300

No.	Site Name	Bird/ Toad	Back	Perf	Comments
8	King's Ruin	toad		loup	Burial 34 (on pumice)
9	King's Ruin	toad	shell	yes	with burial

Middle Verde area: A.D. 1150–1450

No.	Site Name	Bird/ Toad	Back	Perf	Comments
10	Limestone Ruin	toad	shell	yes	Simmons, n. d.; Near head of adult burial; 2 handled B/w; 1 bird B/w; painted wooden plaque; 100s of bone and Ollivella beads
11	Iron Rock Ruin	toad	shell	?	Simmons, n. d.; Burial with B/w
12	Sugarloaf Ruin	toad	shell	yes	Simmons, n. d.
13	Oak Creek	bird	lac?	yes	Blake 1900; John Love collection
14	Bridgeport	toad	shell	?	Simmons, n. d.; near head of burial in room
15–16	Bridgeport	2 toads	shell	?	Simmons, n. d.; in burials
17	Bridgeport	toad	shell	?	Simmons, n. d.; burial; yellowware dipper
18	Bridgeport	toad	shell	?	Simmons, n. d.; adult burial; 2 yellowware bowls
19	Bridgeport	bird	shell	yes	Simmons, n. d.; burial
20–22	Tuzigoot	3 toads	shell	one	Simmons, n. d.; Caywood and Spicer 1935; male priest burial; 2 toads by ears, one on chest as pendant; 1 with notch at base; Jeddito B/y and much wealth
23	Tuzigoot	toad	?	no	Display caption; tiny; with medicine bag
24	Tuzigoot	toad	shell	no	Display caption; small
25	Tuzigoot	Rect.	?		Display caption; with tiny infant

Middle Verde area, continued

No.	Site Name	Bird/ Toad	Back	Perf	Comments
26	Montezuma Castle	bird	Glyc	yes	Jackson 1954; cist grave 3, at ears & under chin of 30–40 yrs female
27	Montezuma Castle	circle	wood	?	on baby's breast

Perry Mesa: A.D. 1300–1450

No.	Site Name	Bird/ Toad	Back	Perf	Comments
28	Pueblo Pato; NA11,434	toad	shell	?	in sealed niche of pueblo room with infant

Phoenix Basin: A.D. 1250–1450

No.	Site Name	Bird/ Toad	Back	Perf	Comments
29	Phoenix Basin?	toad	?	?	collection of Lincoln Fowler
30	Los Muretos	toad	shell	No	in Gila Poly jar with asbestos-like substance
31–33	Casa Grande	toad & 2 birds	toad shell; birds wood		in cache, Compound A, Room E (Huffman 1925); 1 bird and toad large
34	La Ciudad	bird	shell	?	breast of burial near Mound A
35	Los Hornos	toad	?	?	under chin of burial with painted face; much wealth
36	Mesa Grande	toad	shell	?	with infant burial below Rooms A-C, by head and left shoulder; Gila Poly
37	Phoenix Basin	bird	?	?	Jacka and Hammack 1975
38	?	toad	shell	yes	(in Heard Museum); notch in bottom

Gila Bend area: A.D. 1150–1300

No.	Site Name	Bird/ Toad	Back	Perf	Comments
39	Gila Bend area	toad	shell	?	north of 12-mile site

Chaves Pass area: A.D. 1250–1450

No.	Site Name	Bird/ Toad	Back	Perf	Comments
40	Pollack	bird	wood	?	
41	Chaves Pass	toad	shell	?	on breast of burial; possible anus shown (Fewkes 1898)
42	Chaves Pass	toad	shell	yes	

Grasshopper/Kinishba/Gila Pueblo areas: A.D. 1300–1450

No.	Site Name	Bird/ Toad	Back	Perf	Comments
43	Grasshopper	toad	shell		with burial
44	Kinishba	toad	shell	?	next to wall of Room 43, east side of large court; breast ornament?
45	Gila Pueblo	toad	shell	yes	*Arizona Republican* 1925

Rye Creek and Tonto Basin areas: A.D. 1300–1450

No.	Site Name	Bird/ Toad	Back	Perf	Comments
46	Rye Creek area	toad	shell	?	Rim of turquoise beads
47	Keystone Ruin (Roosevelt Lake)	toad	shell	yes	Notch at base
48	U:3:49 ASU	toad	shell	?	Gila phase; on floor of Room 3

Rye Creek and Tonto Basin areas, continued

No.	Site Name	Bird/ Toad	Back	Perf	Comments
49	?	toad	shell	?	(in Mesa SW Museum); notch in base
50	Salado ruin	toad	?	?	Jacka and Hammack 1975; notch in base
51	Salado or Hohokam	toad			"
52	Salado or Hohokam	toad			"
53	Salado or Hohokam	toad	shell		" : on bracelet
54	Tonto Basin	toad	stone	Yes	" (with stripe down back); lignite with turquoise eyes
55	Salado	toad	?	?	"
56	?	bird	?	?	" (end plate)

Southeast Arizona and Casas Grandes

No.	Site Name	Bird/ Toad	Back	Perf	Comments
57	Webb Ruin (Graham Mts)	toad	wood?	?	in room (Duffen 1937)
58	Casas Grandes	bird			(Di Peso, Rinaldo and Fenner 1974:6:459)

References

Abbott, C. C.

1863 Description of a Collection of Jasper Lance Heads Found Near Trenton, New Jersey and Remarks on the Locality with Reference to Indian Antiquities. *Academy of Natural Sciences Proceedings* 15:278.

1872 The Stone Age in New Jersey. *American Naturalist* 6:144–60, 199–229.

1873 Occurrence of Implements in the River Drift of Trenton, New Jersey. *American Naturalist* 7:204–9.

1876a Antiquity of Man. *American Naturalist* 10:51–52.

1876b Indications of the Antiquity of the Indians of North America, Derived from a Study of Their Relics. *American Naturalist* 10:65–72.

1876c Traces of an American Autochthon. *American Naturalist* 10:329–35.

1876d The Stone Age in New Jersey. *Smithsonian Institution Annual Report for 1875,* pp. 246–380.

1877a On the Discovery of Supposed Paleolithic Implements from the Glacial Drift in the Valley of the Delaware River, Near Trenton, New Jersey. *Harvard University, Peabody Museum Annual Report* 10:30–43.

1877b The Classification of Stone Implements. *American Naturalist* 11:495–96.

1881a Historical Sketch of Their Discovery. *Proceedings of the Boston Society of Natural History* 21:124–32.

1881b *Primitive Industry.* G. Bates, Salem, Massachusetts.

1883 Paleolithic Man in Ohio. *Science* 1:359.

1884 *A Naturalist's Rambles About Home.* D. Appleton, New York.

1886a Sketch of Frederic Ward Putnam. *Popular Science Monthly* 29:693–97.

1886b *Upland and Meadow.* Harper and Brothers, New York.

1890 Report of the Curator of the Museum of American Archaeology. *University of Pennsylvania Annual Report* 1:1–54.

1892a Paleolithic Man in America. *Science* 20:270–71.

1892b Paleolithic Man: A Last Word. *Science* 20:344–45.

1892c Recent Archaeological Explorations in the Valley of the Delaware. *University of Pennsylvania Series in Philology, Literature and Archaeology* 2(1):1–30.

1893 The So-called "Cache Implements." *Science* 21:122–23.
1897 Review of "Researches Upon the Antiquity of Man in the Delaware Valley and the Eastern United States," by H. C. Mercer. *The Critic* 796:350.

Adams, H.
1918 [1907] *The Education of Henry Adams.* Random House, New York.

Aiello, L.
1967 Charles Conrad Abbott, M. D. Pick and Shovel Scientist. *New Jersey History* 85:208–16.

American Anthropologist
1900 In Memoriam: Frank Hamilton Cushing. *American Anthropologist* 2(2): 354–80.

Anderson, B.
1983 *Imagined Communities: Reflections on the Origin and Spread of Nationalism.* Verso, London.

Andresen, J. T.
1990 *Linguistics in America, 1769–1924: A Critical History.* Routledge, London.

Anonymous
1882 Abbott's "Primitive Industry." *The Nation* 34:37–39.
1886 The Abbott collection at the Peabody Museum. *Science* 7:4–5.
1887 Sketch of Charles C. Abbott. *Popular Science Monthly* 30:547–53.
1900 In Memoriam. Frank Hamilton Cushing. *Bulletin of the Free Museum of Science and Art* 2(4):257–58.
1907 John E. W. Keely. *The National Cyclopaedia of American Biography* 9:137–38. James T. White, New York.
1916 Work in the Field. *El Palacio* 3(4):84–86.
1916–17 Excavations at Otowi, New Mexico. *Report of the Philadelphia Museums, Commercial Museum, 1916–17.*
1917 In the Field. *El Palacio* 4(3).
1918 Crescencio Martinez. *El Palacio* 5(8):133.
1919 Museum Convention. *El Palacio* 6(13):218.
1942 Milo George Miller (obituary). *American Medical Journal* July 8, 1942.

Archaeological Association of University of Pennsylvania (AAUP)
1890 Circular. University of Pennsylvania Museum Archives.

Arizona Republican

1925 Lore of Ancient Race is Typified in Sacred Frogs Found at Ruin near Globe. *Arizona Republican* January 3, 1925, p. 5, col. 2.

Atwater, C.

1820 Description of the Antiquities Discovered in the State of Ohio and other Western States. *Archaeologia Americana, Transactions of the American Antiquarian Society* 1:105–267.

Babcock, B. A., and N.J. Parezo (editors)

1988 *Daughters of the Desert. Women Anthropologists and the Native American Southwest, 1880–1980.* University of New Mexico Press, Albuquerque.

Baker, L. D.

1998 *From Savage to Negro: Anthropology and the Construction of Race, 1896–1954.* University of California Press, Berkeley.

Baltzell, E. D.

1958 *Philadelphia Gentlemen: The Making of a National Upper Class.* Free Press, Glencoe, Illinois.

1966 *The Protestant Establishment: Aristocracy and Caste in America.* Vintage, New York.

1979 *Puritan Boston and Quaker Philadelphia.* Vintage, New York.

Bandelier, A. F.

1892 *Final Report of Investigations among the Indians of the Southwestern United States, Carried on Mainly in the Years from 1880 to 1885.* Part II. Papers of the Archaeological Institute of America, American Series IV. John Wilson and Sons, Cambridge.

Banforth, D. B., and A. C. Spaulding

1982 Human Behavior, Explanation, Archaeology, History, and Science. *Journal of Anthropological Archaeology* 1:183–85.

Barber, L.

1980 *The Heyday of Natural History, 1820–1870.* Doubleday, Garden City.

Barkan, E.

1992 *The Retreat of Scientific Racism. Changing Conceptions of Race in Britain and the United States between the World Wars.* Cambridge University Press, Cambridge.

Barnes, B., D. Bloor, and J. Henry
1996 *Scientific Knowledge.* University of Chicago Press, Chicago.

Barton, B. S.
1797 *New Views on the Origins of the Tribes and Nations of America.* Printed for the author by John Bioren, Philadelphia.
1799 Observations and Conjectures Concerning Certain Articles which were taken out of an Ancient Tumulus, or Grave at Cincinnati, County of Hamilton, and Territory of the United States, North-West of the River Ohio. *American Philosophical Society Transactions* 4:181–215.
1809 Hints on the Etymology of Certain English Words, and Their Affinity to Words in the Languages of Different European, Asiatic, and American (Indian) Nations. *American Philosophical Society Transactions* 6:145–57.

Bechler, J. L., and F. W. King
1979 *The Audubon Society Field Guide to North American Reptiles and Amphibians.* Alfred A. Knopf, New York.

Bell, W. J., Jr.
1997 *Patriot-Improvers. Biographical Sketches of Members of the American Philosophical Society, 1743–1768.* American Philosophical Society Memoir vol. 226. Philadelphia.

Bender, T.
1993 *Intellect and Public Life: Essays on the Social History of Academic Intellectuals in the United States.* Johns Hopkins University Press, Baltimore.

Billideau, J.
1986 On the Trail of the Turquoise Toad. MS on file, Museum of Northern Arizona Library, Flagstaff.

Blake, W. P.
1900 Mosaics of Chalchuite. *American Antiquarian* 22(2):108–10.

Blumenbach, J. F.
1865 [1786–95] *The Anthropological Treatises of Johann Friederich Blumenbach.* T. Bendyshe, trans. Anthropological Society of London. Longman, Green, London.

Boas, F.
1893 Ethnology at the Exposition. *Cosmopolitan* 15:607–9.
1896 Limitations of the Comparative Method. *Science* 4:901–8.

1904 The History of Anthropology. *Science* 20:513–24.

1911 Introduction. Handbook of American Indian Languages. *Bureau of American*
–22 *Ethnology Bulletin* 40:1–83.

Bohman, N. A. (editor)

1949 *Svenska män och Kvinnor, Biografisk Uppslagsbok* 6. Albert Bonniers Förlag, Stockholm.

Boston Evening Transcript

1916 Our Ancient Egypt Here at Home. August 30, Part II, p. 4.

Bourne, R. S.

1964 *War and the Intellectuals, Collected Essays, 1915–1919,* edited by C. Resek. Harper Torchbooks, New York.

Bozeman, T. D.

1977 *Protestants in an Age of Science.* University of North Carolina Press, Chapel Hill.

Brackenridge, H. H.

1818 On the Population and Tumuli of the Aborigines of North America, in a Letter to Thomas Jefferson. *American Philosophical Society Transactions,* n.s. 1:151–60.

Brinton, D. G.

1886 Anthropology and Ethnology. *The Iconographic Encyclopedia* 1:17–18.

1890a *Essays of an Americanist.* Porter and Coats, Philadelphia.

1890b *Races and Peoples: Lectures on the Science of Ethnography.* N. D. C. Hodges, New York.

1891 *The American Race: Linguistic Classification and Ethnographic Description of the Native Tribes of North and South America.* N. D. C. Hodges, New York.

1892a *Anthropology as a Science and as a Branch of University Education.* Privately printed, Philadelphia.

1892b On Quarry Rejects. *Science* 20:260–61.

1892c The Nomenclature and Teaching of Anthropology (with comment by J. W. Powell). *American Anthropologist,* o.s. 5: 263–71.

1895 The Aims of Anthropology. *Proceedings of the American Association for the Advancement of Science* 44:1–17.

1902 *The Basis of Social Relations: A Study in Ethnic Psychology,* edited by Livingston Farrand. Putnam's Sons, New York.

Brew, J. O.
1966 *Early Days of the Peabody Museum at Harvard University.* Harvard University Press, Cambridge.

Brody, J. J.
1997 *Pueblo Indian Painting: Tradition and Modernism in New Mexico, 1900–1930.* School of American Research, Santa Fe.

Bronner, S. J.
1989 Object Lessons: The Work of Ethnological Museum and Collections. In *Consuming Visions, Accumulation and Display of Goods in America, 1880–1920,* edited by S. J. Bronner, pp. 217–54. W. W. Norton, New York.

Brooks, V. W.
1915 *America's Coming of Age.* B. W. Huebsch, New York.
1952 *The Confident Years, 1885–1915.* E. P. Dutton, New York.

Brown, J. H.
1903 *Lamb's Biographical Dictionary of the United States,* vol. 5. Federal Book, Boston.

Burt, N.
1963 *The Perennial Philadelphians: The Anatomy of an American Aristocracy.* University of Pennsylvania Press, Philadelphia.

Butzer, K.
1982 *Archaeology as Human Ecology.* University of Chicago Press, Chicago.

Carter, E. C., II
1993 *"One Grand Pursuit": A Brief History of the American Philosophical Society's First 250 Years, 1743–1993.* American Philosophical Society, Philadelphia.

Caywood, L. R., and E. H. Spicer
1935 Tuzigoot: The Excavation and Repair of a Ruin on the Verde River near Clarkdale, Arizona (mimeographed). Southwestern Monuments Association, Coolidge, Arizona.

Cheyney, E. P.
1940 *History of the University of Pennsylvania, 1740–1940.* University of Pennsylvania Press, Philadelphia.

Clarke, D.
1973 Archaeology: the Loss of Innocence. *Antiquity* 47:6–18.

Clavigero, F. J.
1979 [1787] *The History of Mexico.* 2 vols. Garland, New York.

Cochran, T. C., and W. Miller
1961 *The Age of Enterprise, A Social History of Industrial America.* Harper and Row, New York.

Cole, D.
1985 *Captured Heritage, the Scramble for Northwest Coast Artifacts.* University of Oklahoma Press, Norman.
1999 *Franz Boas: The Early Years, 1858–1906.* University of Washington Press, Seattle.

Collier, D., and H. Tschopick Jr.
1954 The Role of Museums in American Anthropology. *American Anthropologist* 56:768–79.

Collins, R.
1998 *The Sociology of Philosophies. A Global Theory of Intellectual Change.* Belknap Press of Harvard University Press, Cambridge.

Conn, S.
1998 *Museums and American Intellectual Life, 1876–1926.* University of Chicago Press, Chicago.

Cotter, J. L.
1911–1999 Cotter Papers. University of Pennsylvania Museum of Archaeology and Anthropology, Division of Archives, Philadelphia.
1937 The Occurrence of Flints and Extinct Animals in Pluvial Deposits near Clovis, New Mexico. Part IV: Report on Excavation at the Gravel Pit. *Academy of Natural Sciences Proceedings* 90:2–16.
1937 The Occurrence of Flints and Extinct Animals in Pluvial Deposits near Clovis, New Mexico, part VI: Report on the Field Season of 1937. *Academy of Natural Sciences Proceedings* 90:113–17.
1958 *Archaeological Excavations at Jamestown, Virginia.* National Park Service, Archaeological Research Series 4. Washington, D.C.

Cotter, J. L., D. G. Roberts, and M. Parrington
1992 *The Buried Past: An Archaeological History of Philadelphia.* University of Pennsylvania Press, Philadelphia.

Croll, J.
1875 *Climate and Time in Their Geological Relations: a Theory of Secular Changes of the Earth's Climate.* E. Stanford, London.

Culin, S.
1907 Games of the North American Indians. *Twenty-fourth Annual Report of the Bureau of American Ethnology for the Years 1902–1903:* 1–846. Washington, D.C.
1992 *Games of the North American Indians.* Introduction by D. Tedlock. University of Nebraska Press, Lincoln.

Cushing, F. H.
1893 Diaries for 1893, transcribed by D. R. Wilcox, Museum of Northern Arizona. On file, National Anthropological Archives, Washington D.C.
1895 The Arrow. *American Anthropologist,* o.s. 8(4):307–49.
1897 Exploration of Ancient Key-Dweller Remains on the Gulf Coast of Florida. *Proceedings of the American Philosophical Society* 35(153).

Daniel, G.
1976 *A Hundred and Fifty Years of Prehistory.* Harvard University Press, Cambridge.

Daniels, A. K.
1988 *Invisible Careers: Women Civic Leaders from the Volunteer World.* University of Chicago Press, Chicago.

Darnell, R.
1970 The Emergence of Academic Anthropology at the University of Pennsylvania. *Journal of the History of the Behavioral Sciences* 6:80–92.
1988 *Daniel Garrison Brinton: "The Fearless Critic" of Philadelphia.* University of Pennsylvania Publications in Anthropology 3.
1998 *And Along Came Boas: Continuity and Revolution in Americanist Anthropology.* John Benjamins, Amsterdam.
2001 *Invisible Genealogies. A History of Americanist Anthropology.* University of Nebraska Press, Lincoln.

Darrah, W. C.
1951 *Powell of the Colorado.* Princeton University Press, Princeton.

Davis, M. B.
1987 Field Notes of Clarence B. Moore's Southeast Archaeological Expeditions, 1891–1918: A Guide to the Microfilm Edition. Huntington Free Library, Bronx, New York.

Davis, W., and A. T. Weil
1992 Identity of a New World Psychoactive Toad. *Ancient Mesoamerica* 3:51–59.

Dawkins, W. B.
1883 Early Man in America. *North American Review* 137:338–49.

Deacon, D.
1997 *Elsie Clews Parsons, Inventing Modern Life.* University of Chicago Press, Chicago.

De Laguna, F. (editor)
1960 *Selected Papers from the American Anthropologist, 1898–1920.* Row, Peterson, Evanston.

Dexter, R. W.
1966 Putnam's Problems Popularizing Anthropology. *American Scientist* 54(3): 315–32.

Dieserud, J.
1908 *Historical Review, Library Classification and Select Annotated Bibliography; with a list of the Chief Publications of Leading Anthropological Societies and Museums.* Open Court, Chicago.

Di Peso, C. C., J. B. Rinaldo, and G. Fenner
1974 *Casas Grandes, A Fallen Trading Center of the Gran Chichimeca,* vol. 6. Amerind Foundation no. 9. Dragoon, Arizona.

Dockstader, F. J.
1967 Anthropology and the Museum. In *The Philadelphia Anthropological Society: Papers Presented on its Golden Anniversary,* edited by J. W. Gruber, pp. 132–42. Temple University Press, Philadelphia.

Duffen, W. A.

1937 Some Notes on a Summer's Work Near Bonita, Arizona. *Kiva* 2(4):13–16.

Duncan, F.

1882 "An Estate Twice Saved by a Mother." 4 pg. disbound from unknown source. Academy of Natural Sciences of Philadelphia, Coll. 375: Jessup, Augustus Edward, 1799–1859. Papers, 1822–1913.

Dyke, L. F.

1989 Henry Chapman Mercer: an Annotated Chronology.*The Mercer Mosaic: Journal of the Bucks County Historical Society* 6(2–3):34–67.

Dyson, S. L.

1998 *Ancient Marbles to American Shores. Classical Archaeology in the United States.* University of Pennsylvania Press, Philadelphia.

1999 Brahmins and Bureaucrats: Some Reflections on the History of American Classical Archaeology. *In Assembling the Past: Studies in the Professionalization of Archaeology,* edited by A. B. Kehoe and M. B. Emmerich, pp. 103–16. University of New Mexico Press, Albuquerque.

Ennis, J.

1866 On Kitchen Middens Near Cape May Court House, New Jersey. *Academy of Natural Sciences Proceedings* 18:290–92.

Evans, B.

1997 Cushing's Zuni Sketchbooks: Literature, Anthropology, and American Notions of Culture. *American Quarterly* 49(4):717–45.

Evans, J.

1872 *The Ancient Stone Implements, Weapons and Ornaments of Great Britain.* Longmans, London.

Evening Bulletin, Philadelphia

1916 "Phila. Explorers Murdered, Report." April 20.

1916 "Broad St. Speeders Fear of Explorers." May 25.

Evening Times Union, Jacksonville

1895 "To Hunt For Skulls," August 28:1.

Ewan, J.

1986 One Professor's Chief Joy: A Catalog of Books Belonging to Benjamin Smith Barton. In *Science and Society in Early America, Essays in Honor of*

Whitfield J. Bell, Jr., edited by R. S. Klein, pp. 311–44. American Philosophical Society Memoir vol. 166. Philadelphia.

Ewers, J. C.
1979 *Indian Art in Pipestone: George Catlin's Portfolio in the British Museum.* Smithsonian Institution Press, Washington, D.C.

Fane, D.
1991 The Language of Things: Stewart Culin as Collector. In *Objects of Myth and Memory: American Indian Art at the Brooklyn Museum,* edited by D. Fane, I. Jacknis, and L. M. Breen, pp. 13–27. Brooklyn Museum, Brooklyn, New York.

Fewkes, J. W.
1898 Preliminary Account of an Expedition to the Pueblo Ruins Near Winslow, Arizona, in 1896. *Annual Report of the Smithsonian Institution for 1896,* pp. 517–40. Washington, D.C.

Finkel, K.
1980 *Nineteenth-Century Photography in Philadelphia.* Dover, New York.

Florida Times Union
1895 "Moore Probes the Mounds." *Florida Times Union* November 28:7, Jacksonville.

Fowler, D. D.
1999 Harvard vs. Hewett. The Contest for Control of Southwestern Archaeology, 1904–1930. In *Assembling the Past. Studies in the Professionalization of Archaeology,* edited by A. B. Kehoe and M. B. Emmerichs, pp. 165–211. University of New Mexico Press, Albuquerque.
2000 *A Laboratory for Anthropology. Science and Romanticism in the American Southwest, 1846–1930.* University of New Mexico Press, Albuquerque.
2003 "Natural History of Man." Reflections on Anthropology, Museums and Science. In *A History of Anthropology at the Field Museum, 1890–2000,* edited by S. E. Nash and G. Feinman. Field Museum of Natural History, Chicago, in press.

Fowler, D. D., and M. W. Salter
2004 Archaeological Ethics in Context and Practice. In Handbook of Archaeological Theories, edited by C. Chippindale and H. Maschner. AltaMira Press, Walnut Creek, California, in press.

Fowler, D. D., and D. R. Wilcox
1999 From Thomas Jefferson to the Pecos Conference: Changing Anthropological Agendas in the North American Southwest. In *Surveying the Record: North American Scientific Exploration to 1930,* edited by E. C. Carter II, pp. 197–223. American Philosophical Society Memoirs 231. Philadelphia.

Frazer, P.
1897 Obituary of Edward Drinker Cope. *American Naturalist* 31(365):410–11.

Fuller, W. P.
1972 *St. Petersburg and Its People.* Great Outdoors Publishing Co., St. Petersburg, Florida.

Gabb, W. M.
1868 On Kitchen Middens Near San Francisco. *Academy of Natural Sciences Proceedings* 20:182.

Gallatin, A.
1836 A Synopsis of the Indian Tribes of North America. *Archaeologia Americana, Transactions of the American Antiquarian Society* 2:9–422.

Geikie, J.
1874 *The Great Ice Age and Its Relation to the Antiquity of Man.* Daldy, Isbister, London.

Gerbi, A.
1973 *The Dispute of the New World: The History of a Polemic, 1750–1900,* rev. ed. University of Pittsburgh Press, Pittsburgh.

Gilliland, M. S.
1975 *The Material Culture of Key Marco, Florida.* University of Florida Press, Gainesville.
1989 *Key Marco's Buried Treasure, Archaeology and Adventure in the Nineteenth Century.* University of Florida Press, Gainesville.

Gliddon, G. R.
1842 On Egyptian Remains. *Academy of Natural Sciences Proceedings* 1:172–74.

Goddard, I.
1996 The Description of the Native Languages of North America before Boas. *Handbook of North American Indians* 17:17–42. Smithsonian Institution, Washington, D.C.

Goetzmann, W. H.
1966 *Exploration and Empire: The Explorer and the Scientist in the Winning of the American West.* Alfred A. Knopf, New York.

Goggin, J. M.
1952 *Space and Time Perspectives in Northern St. Johns Archeology, Florida.* Yale University Publications in Anthropology 47, New Haven.

Goggin, J. M., and W. C. Sturtevant
1964 The Calusa, A Stratified Nonagricultural Society (with notes on Sibling Marriage). In *Explorations in Cultural Anthropology in Honor of George Peter Murdock,* edited by W. H. Goodenough, pp. 179–219. McGraw-Hill, New York.

Goodrich, Samuel G.
1848 *History of the Indians of North and South America by the Author of Peter Parley's Tales.* Ravel and Mann, Boston.

Green, C. M.
1962 *Washington, A History of the Capital, 1800–1950.* Princeton University Press, Princeton.

Green, J. (editor)
1979 *Zuñi: Selected Writings of Frank Hamilton Cushing.* University of Nebraska Press, Lincoln.
1990 *Cushing at Zuñi: The Correspondence and Journals of Frank Hamilton Cushing, 1879–1884.* University of New Mexico Press, Albuquerque.

Gruber, J. W.
1967 Preface. In *The Philadelphia Anthropological Society: Papers Presented on its Golden Anniversary,* edited by J. W. Gruber, pp. v–ix. Temple University Press, Philadelphia.

Hallowell, A. I.
1960 The Beginnings of Anthropology in America. In *Selected Papers from the American Anthropologist, 1888–1920,* edited by F. De Laguna, pp. 1–104. Row, Peterson, Evanston.

Hardesty, D. L., and D. D. Fowler
2001 Archaeology and Environmental Changes. In *New Directions in Anthropology and Environment: Intersections,* edited by C. L. Crumley, pp. 72–89. AltaMira Press, Walnut Creek, California.

Harrington, J. C.
1952 Historic Site Archaeology in the United States. In *Archaeology of the Eastern United States,* edited by J. B. Griffin, pp. 335–44. University of Chicago Press, Chicago.
1953 Archaeology and Local History. *Indiana Magazine of History* 49(1):69–79.
1955 Archaeology as an Auxiliary Science to American History. *American Anthropologist* 57(6):1121–30.

Harrington, J. P.
1916 Ethnogeography of the Tewa Indians. *Twenty-ninth Annual Report of the Bureau of American Ethnology, 1907–1908,* pp. 29–618.

Harris, N.
1962 The Gilded Age Revisited: Boston and the Museum Movement. *American Quarterly* 14(4):545–66.
1990 *Cultural Excursions, Marketing Appetites and Cultural Tastes in Modern America.* University of Chicago Press, Chicago.

Haury, E. W.
1945 *The Excavations of Los Muertos and Neighboring Ruins in the Salt River Valley, Southern Arizona.* Papers of the Peabody Museum of American Archaeology and Ethnology 24(1). Peabody Museum, Cambridge.
1995 Wherefore a Harvard Ph.D.? *Journal of the Southwest* 37(4):710–33.

Hayden, F. V.
1866 On the Pipe-stone Quarry of Northeast Dakota. *Academy of Natural Sciences Proceedings* 18:291.

Hewett, E. L.
1904 Archaeology of the Pajarito Park, New Mexico. *American Anthropologist* 6:629–59.
1906 *Antiquities of the Jemez Plateau, New Mexico.* Bureau of American Ethnology Bulletin no. 32.
1918 Crescencio Martinez, Artist. *El Palacio* 5(5):67–69.
1938 *Pajarito Plateau and Its Ancient People.* University of New Mexico Press and School of American Research, Albuquerque.

Hinsley, C. M.
1979 Anthropology as Science and Politics: The Dilemmas of the Bureau of American Ethnology, 1879 to 1904. In *The Uses of Anthropology,* edited by

W. Goldschmidt, pp. 15–32. Publication of the American Anthropological Association no. 11. Washington, D.C.

1981 *Savages and Scientists: The Smithsonian Institution and the Development of American Anthropology, 1846–1910.* Smithsonian Institution Press, Washington, D.C.

1985 From Shell-Heaps to Stelae: Early Anthropology at the Peabody Museum. In *Objects and Others: Essays on Museums and Material Culture,* edited by G. W. Stocking Jr., pp. 49–74. University of Wisconsin Press, Madison.

1983 Ethnographic Charisma and Scientific Routine: Cushing and Fewkes in the American Southwest, 1879–1893. In *Observers Observed: Essays on Ethnographic Fieldwork,* edited by G. W. Stocking Jr., pp. 53–69. University of Wisconsin Press, Madison.

1989 Revising and Revisioning the History of Archaeology. In *Tracing Archaeology's Past: The Historiography of Archaeology,* edited by A. L. Christenson, pp. 79–96. Southern Illinois University Press, Carbondale.

1992 The Museum Origins of Harvard Anthropology, 1866–1915. In *Science at Harvard University: Historical Perspectives,* edited by C. A. Elliott and M. W. Rossiter, pp. 121–45. Lehigh University Press, Bethlehem, Pennsylvania.

1993 In Search of the New World Classical. In *Collecting the Pre-Columbian Past,* edited by E. Boone, pp. 105–21. Dumbarton Oaks, Washington, D.C.

1999 Frederic Ward Putnam (1839–1915). In *Encyclopedia of Archaeology: The Great Archaeologists,* edited by T. Murray, pp. 141–54. ABC-Clio, Santa Barbara.

Hinsley, C. M., and B. Holm

1976 A Cannibal in the National Museum: The Early Career of Franz Boas in America. *American Anthropologist* 78: 306–16.

Hinsley, C. M., and D. R. Wilcox (editors)

1995 A Hemenway Portfolio. *Journal of the Southwest* 37(4):517–744.

Hodge, F. W.

1945 Foreword. In *The Excavation of Los Muertos and Neighboring Ruins in the Salt River Valley, Southern Arizona,* by E. W. Haury, pp. iv–v. Papers of the Peabody Museum of American Archaeology and Ethnology, Harvard University 24(1).

Hodge, F. W. (editor)

1907–10 *Handbook of the Indians North of Mexico,* 2 vols. Bureau of American Ethnology Bulletin 30.

Holmes, W. H.
1890 A Quarry Workshop of the Flaked Stone Implement Makers in the District of Columbia. *American Anthropologist,* o.s. 3:1–26.
1892 Modern Quarry Refuse and the Paleolithic Theory. *Science* 20:295–97.
1893a Distribution of Stone Implements in the Tidewater Country. *American Anthropologist,* o.s. 6:1–14.
1893b Are There Traces of Man in the Trenton Gravels? *Journal of Geology* 1:15–37.
1893c Glacial Man and Paleolithic Culture: a Preliminary Word. *Science* 21:29–30.
1893d Traces of Glacial Man in Ohio. *Journal of Geology* 1:147–63.
1893e A Question of Evidence. *Science* 21:135–36.
1893f Vestiges of Early Man in Minnesota. *American Geologist* 11:219–40.
1894 Earthenware of Florida, Collections of Clarence B. Moore. *Journal of the Academy of Natural Sciences of Philadelphia* 10:105–28.
1903 Aboriginal Pottery of the Eastern United States. *Twentieth Annual Report of the Bureau of American Ethnology, 1898–1899,* pp. 1–201.
1916 In Memoriam: Matilda Coxe Stevenson. *American Anthropologist* 18(4): 552–59.

Huddleston, L. E.
1967 *Origins of the American Indians. European Concepts, 1492–1729.* University of Texas Press, Austin.

Huffman, J. W.
1925 Turquoise Mosaics from Casa Grande. *Art and Archaeology* 20(2):82–88.

Hughes, R.
1997 *American Visions, The Epic History of Art in America.* Alfred A. Knopf, New York.

Jacka, J. J., and N. S. Hammack
1975 *Indian Jewelry of the Prehistoric Southwest.* University of Arizona Press, Tucson.

Jacknis, I.
2000 A Museum Prehistory: Phoebe Hearst and the Founding of the Museum of Anthropology, 1891–1901. *Chronicle of the University of California* 4:47–77.

Jackson, E., and S. P. Van Valkenburg
1954 *Montezuma Castle Archaeology, Part 1: Excavations.* Southwestern Monuments Association Technical Series 3, Part 1. Globe, Arizona.

James, E. T. (editor)
1971 *Notable American Women 1607–1950: A Biographical Dictionary,* 2 vols. Harvard University Press, Cambridge.

James, H.
1967 [1905] *The American Scene.* Horizon Press, New York.

Jefferson, T.
1944 Notes on Virginia. In *The Life and Selected Writings of Thomas Jefferson,* edited by A. Koch and W. Peden, pp. 185–288. Modern Library, New York.

Jefferson, T., et al
1799 Circular Letter. *American Philosophical Society Transactions* 4: xxxvii–xxxix. Reprinted 1809 in Transactions, vol. 5.

Jernigan, E. W.
1978 *Jewelry of the Prehistoric Southwest.* School of American Research, Santa Fe.

Joha, J.
1992 Unpublished Paper Presented at the Southwest Symposium, January 1992. University of Arizona, Tucson.

Johns, E.
1983 *Thomas Eakins. The Heroism of Modern Life.* Princeton University Press, Princeton.

Joiner, C.
1992 The Boys and Girls of Summer: The University of New Mexico Archaeological Field School in Chaco Canyon. *Journal of Anthropological Research* 48:49–66.

Jordan, J. W. (editor)
1911 *Colonial Families in Philadelphia,* 2 vols. Lewis, Philadelphia.

Josephy, A. M., Jr. (editor)
1992 *America in 1492. The World of the Indian Peoples Before the Arrival of Columbus.* Alfred A. Knopf, New York.

Judd, N. M., M. R. Harrington, and S. K. Lothrop
1957 Frederick Webb Hodge—1864–1956. *American Antiquity* 22(4):401–4.

Kehoe, A. B.
1995 Scientific Creationism: World View, Not Science. In *Cult Archaeology and Creationism*, 2nd ed., edited by F. B. Harrold and R. A. Eve, pp. 11–20. University of Iowa Press, Iowa City.
1998 *The Land of Prehistory: A Critical History of American Archaeology*. Routledge, New York.
1999a Introduction. In *Assembling the Past: Studies in the Professionalization of Archaeology*, edited by A. B. Kehoe and M. B. Emmerichs, pp. 1–18. University of New Mexico Press, Albuquerque.
1999b Professionals May Not Be Women, pp. 117–19. In *Assembling the Past: Studies in the Professionalization of Archaeology*, edited by A. B. Kehoe and M. B. Emmerichs, pp. 117–19. University of New Mexico Press, Albuquerque.

Kehoe, A. B., and M. B. Emmerichs (editors)
1999 *Assembling the Past: Studies in the Professionalization of Archaeology*. University of New Mexico Press, Albuquerque.

Kelly, M. W.
2000 The Yellowstone Expedition: The American Military on the Missouri River 1818–1820. A Tale of Politics, Personalities, and Pettiness. *Journal of America's Military Past* 27(1):17–53.

Kennedy, P.
1987 *The Rise and Fall of the Great Powers*. Random House, New York.

Kennedy, R. G.
1994 *Hidden Cities. The Discovery and Loss of Ancient North American Civilization*. Penguin, New York.

King, E. M., and B. P. Little
1986 George Byron Gordon and the Early Development of The University Museum. In *Raven's Journey: The World of Alaska's Native People*, edited by S. A. Kaplan and K. J. Barsness, pp. 16–53. University Museum, University of Pennsylvania, Philadelphia.

Kroeber, A. L.
1930–35 Cushing, Frank Hamilton (1857–1900). *Encyclopedia of the Social Sciences* 4:457.
1954 The Place of Anthropology in Universities. *American Anthropologist* 56: 764–65.

Kuklick, B.
1996 *Puritans in Babylon. The Ancient Near East and American Intellectual Life, 1880–1930.* Princeton University Press, Princeton.

Kunz, G. F.
1890 *Gems and Precious Stones of North America.* Scientific Publications, New York.

Lartet, E., and H. Christy
1875 *Reliquiae Aquitanicae. Being Contributions to the Archaeology and Paleontology of Perigord and the Adjoining Provinces of Southern France.* Williams and Norgate, London.

Levine, M. A.
1999 Uncovering a Buried Past. Women in Americanist Archaeology Before the First World War. In *Assembling the Past. Studies in the Professionalization of Archaeology,* edited by A. B. Kehoe and M. B. Emmerichs, pp. 133–51. University of New Mexico Press, Albuquerque.

Levine, L. W.
1988 *Highbrow/Lowbrow: The Emergence of Cultural Hierarchy in America.* Harvard University Press, Cambridge.

Lewis, A., J. Turner, and S. McQuillan
1987 *The Opulent Interiors of the Gilded Age.* Dover, New York.

Lewis, R. B. (editor)
1996 *Kentucky Archaeology.* University Press of Kentucky, Lexington.

Lewis, T. M. N., and M. Kneberg
1970 [1946] *Hiwassee Isand: An Archaeological Account of Four Tennessee Indian Peoples.* University of Tennessee Press, Knoxville.

Lubbock, J.
1865 *Prehistoric Times.* Williams and Norgate, London.

Lyman, S. M.
1982 Two Neglected Pioneers of Civilizational Analysis: Perspectives of R. Stewart Culin and Frank Hamilton Cushing. *Social Research* (autumn 1982): 694–729.

Madeira, P. C., Jr.
1964 *Men in Search of Man. The First Seventy-five Years of the University Museum of the University of Pennsylvania.* University of Pennsylvania Press, Philadelphia.

Mark, J.
1980 *Four Anthropologists: An American Science in its Early Years.* Science History, New York.

Mason, O. T.
1877 Anthropological News. *American Naturalist* 11:308–11.
1882 What is Anthropology? Saturday Lectures to the Anthropological Society of Washington. *Smithsonian Institution Annual Report for 1882,* pp. 633–73.
1883 An Account of the Progress of Anthropology in the Year 1881. *Smithsonian Institution Annual Report for 1881,* pp. 499–525.
1884 The Scope and Value of Anthropological Studies. *Proceedings of the American Association for the Advancement of Science* 2:367–83.
1891 Progress of Anthropology in 1890. *Smithsonian Institution Annual Report for 1890,* pp. 527–608.

Mason, O. T., et al.
1889 The Aborigines of the District of Columbia and the Lower Potomac. *American Anthropologist* 2:225–68.

Marching, L.
1999 Clara Sophia Jessup Bloomfield Moore. *American National Biography* 15: 741–42. Oxford University Press, New York.

Mathien, F. J.
1992 Women of Chaco: Then and Now. In *Rediscovering Our Past: Essays on the History of American Archeology,* edited by J. E. Reyman, pp. 103–30. Avebury Press, Glasgow.

May, H. F.
1964 *The End of American Innocence.* Quadrangle, Chicago.

McGee, W. J.
1888 Paleolithic Man in America: His Antiquity and His Environment. *Popular Science Monthly* 34:20–36.
1897 The Science of Humanity. *Science* 6:413–33.
1905 Anthropology and Its Larger Problems. *Science* 21:770–84.

McGoun, W. E.

1993 *Prehistoric Peoples of South Florida.* University of Alabama Press, Tuscaloosa.

McGregor, J. C.

1943 Burial of an Early American Magician. *Proceedings of the American Philosophical Society* 86(2):270–98.

McVicker, D.

1999 Buying a Curator: Establishing Anthropology at Field Columbian Museum. In *Assembling the Past, Studies in the Professionalization of Archaeology,* edited by A. B. Kehoe and M. B. Emmerichs, pp. 37–52. University of New Mexico Press, Albuquerque.

Meltzer, D. J.

1983 The Antiquity of Man and the Development of American Archaeology. *Advances in Archaeological Method and Theory* 6:1–51.

1991 On "Paradigms" and "Paradigm Bias" in Controversies over Human Antiquity in America. In *The First Americans: Search and Research,* edited by T. Dillehay and D. Meltzer, pp. 13–49. CRC Press, Boca Raton, Florida.

1994 The Discovery of Deep Time: A History of Views on the Peopling of the Americas. In *Method and Theory for Investigating the Peopling of the Americas,* edited by R. Bonnichsen and D. G. Steele, pp. 7–26. Center for the Study of the First Americans, Corvallis, Oregon.

Menard, L.

2001 *The Metaphysical Club. A Story of Ideas in America.* Farrar, Strauss and Giroux, New York.

Mercer, H. C.

1897 Researches Upon the Antiquity of Man in the Delaware Valley. Ginn, Boston.

Meyerson, M., and D. P. Winegrad

1978 *Gladly Learn and Gladly Teach. Franklin and His Heirs at the University of Pennsylvania, 1740–1976.* University of Pennsylvania Press, Philadelphia.

Milanich, J. T., and S. Milbraith (editors)

1989 *First Encounters. Spanish Exploration in the Caribbean and the United States, 1492–1570.* University of Florida Press and Florida Museum of Natural History, Gainesville.

Miller, G. L.
1998 Memorial: J. C. Harrington, 1901–1998. *Historical Archaeology* 32(4):1–7.

Mitchem, J. M.
1999 Introduction: Clarence B. Moore's Research in East Florida, 1873–1896. In *The East Florida Expeditions of Clarence Bloomfield Moore*, edited by J. M. Mitchem, pp. 1–52. University of Alabama Press, Tuscaloosa.

Moore, C. J.
1940 *Ancestry of Clarence Bloomfield Moore, of Philadelphia*, edited by H. de Bildt and M. Rubincam. National Genealogical Society Quarterly (March).

Moore, C. B.
1892a A Burial Mound of Florida. *American Naturalist* 26:129–43.
1892b Certain Shell Heaps of the St. John's River, Florida, hitherto Unexplored (First Paper). *American Naturalist* 26:912–22.
1892c Mounds in Florida. *American Antiquarian and Oriental Journal* 14:292–95.
1892d Supplementary Investigations at Tick Island. *American Naturalist* 26:568–79.
1893a Certain Shell Heaps of the St. John's River, Florida, hitherto Unexplored (Second Paper). *American Naturalist* 27:8–13, 113–17.
1893b Certain Shell Heaps of the St. John's River, Florida, hitherto Unexplored (Third Paper). *American Naturalist* 27:605–24.
1893c Certain Shell Heaps of the St. John's River, Florida, hitherto Unexplored (Fourth Paper). *American Naturalist* 27:708–23.
1894a Certain Sand Mounds of the St. John's River, Florida, Part II. *Journal of the Academy of Natural Sciences of Philadelphia* 10:129–246.
1894b Certain Shell Heaps of the St. John's River, Florida, hitherto Unexplored (Fifth Paper). *American Naturalist* 28:15–26.
1894c Tobacco Pipes in Shell-heaps of the St. John's. *American Naturalist* 28: 622–23.
1895 Certain River Mounds of Duval County, Florida. *Journal of the Academy of Natural Sciences of Philadelphia* 10:448–502.
1902 Archaeological Research in the Southern United States. *Proceedings of the 13th International Congress of Americanists, New York*, pp. 27–40.
1903a Sheet-Copper from the Mounds is Not Necessarily of European Origin. *American Anthropologist* 5:27–49.
1903b The So-called Hoe-Shaped Implement. *American Anthropologist* 5:498–502.
1904 Aboriginal Urn-burial in the United States. *American Anthropologist* 6:660–69.
1905a A Form of Urn-burial on Mobile Bay. *American Anthropologist* 7:167–68.
1907 Notes on the Ten Thousand Islands. *Journal of the Academy of Natural Sciences of Philadelphia* 13:457–70.

1910 Antiquities of the St. Francis, White, and Black Rivers, Arkansas. *Journal of the Academy of Natural Sciences of Philadelphia* 14:253–364.

1914 The Red-Paint People of Maine. *American Anthropologist* 16:137–39.

1915a Aboriginal Sites on Tennessee River. *Journal of the Academy of Natural Sciences of Philadelphia* 16:170–428.

1915b The "Red-Paint People"—II. *American Anthropologist* 17:209.

1918 The Northwestern Florida Coast Revisited. *Journal of the Academy of Natural Sciences of Philadelphia* 16:513–79.

1919 Notes on the Archaeology of Florida. *American Anthropologist* 21:400–402.

1921a Notes on Shell Implements from Florida. *American Anthropologist* 23:12–18.

1921b Shell Implements from Florida. *Academia Nacional de Historia Bolétin* 1:1–3. Quito.

1922 [1896] *Additional Mounds of Duval and of Clay Counties, Florida; Mound Investigation on the East Coast of Florida; Certain Florida Coast Mounds North of the St. Johns River.* Indian Notes and Monographs. Museum of the American Indian, Heye Foundation, New York. Privately printed, Philadelphia.

1987 Field Notes of Clarence B. Moore's Southeastern Archaeological Expeditions, 1891–1918, M. B. Davis (compiler). Microfilm, 6 rolls, Huntington Free Library and Museum of the American Indian Library, New York.

Moore, E.

1970–72 Women We Have Known. Unpublished Manuscript in the Historical Society of Pennsylvania Collection 2138. American Association of University Women, Pennsylvania Division, Philadelphia Branch. Box 2, Executive Director's Office. AAUW History by Ellen Moore, 1970–72[?].

Morgan, L. H.

1877 *Ancient Society.* Henry Holt, New York.

Morton, S. G.

1841 Observations on a Mode of Ascertaining the Internal Capacity of the Human Cranium. *Academy of Natural Sciences Proceedings* 1:7–8.

Mueller, E. A.

1983 *Ocklawaha River Steamboats.* Mendelson, Jacksonville.

Murrowchick, R. E.

1990 A Curious Sort of Yankee: Personal and Professional Notes on Jeffries Wyman (1814–74). *Southeastern Archaeologist* 9:55–66.

New York Times
1934 W. Dinwiddie Dies (Editor and Author). *New York Times*, June 18, 1934, p. 17, col. 3.

Nillson, S.
1868 *Primitive Inhabitants of Scandinavia*, 3rd ed. Edited and with an Introduction by Sir J. Lubbock. Longmans, Green, London.

Noah, M. M.
1837 *Discourse on the Evidence of the American Indians Being Descendants of the Lost Tribes of Israel.* James Van Norden, New York.

Nolan, E. J. (editor)
1913 An Index to the Scientific Contents of the Journal and Proceedings of the Academy of Natural Sciences of Philadelphia. *Academy of Natural Sciences Proceedings* vol. 62: i–xiv, 1–1419.

Norton, C. E.
1885 The First American Classical Archaeologist. *American Journal of Archaeology* 1:3–9.

Nott, J. C., and G. R. Gliddon
1854 *Types of Mankind: or, Ethnological Researches, Based Upon the Ancient Monuments, Paintings, Sculptures, and Crania of Races and Upon Their Natural, Geographical, Philological and Biblical History.* Lippincott, Grambo, Philadelphia.

O'Connor, D., and D. Silverman
1979 The University Museum in Egypt. *Expedition* (winter 1979):4–43.

Orvell, M.
1989 *The Real Thing: Imitation and Authenticity in American Culture, 1880–1940.* University of North Carolina Press, Chapel Hill.

Osgood, C.
1979 *Anthropology in Museums in Canada and the United States.* Milwaukee Public Museum, Milwaukee.

Panzer, M.
1982 *Philadelphia Naturalistic Photography 1865–1906.* Yale University Art Gallery, New Haven.

Paper Trade Journal
1878 Personals Column item about return of Clarence B. Moore. *Paper Trade Journal* September 21:304.
1899 The Jessup and Moore Paper Company Reorganized. *Paper Trade Journal* September 14:492.

Parezo, N.J. (editor)
1993 *Hidden Scholars: Women Anthropologists and the Native American Southwest.* University of New Mexico Press, Albuquerque.

Peabody, G.
1868 Foundation. *First Annual Report of the Trustees of the Peabody Museum of American Archaeology and Ethnology,* pp. 25–28. John Wilson, Cambridge.

Pearson, C. E., T. C. Birchett, and R. A. Weinstein
2000 An Aptly Named Steamboat: Clarence B. Moore's Gopher. *Southeastern Archaeology* 19:82–87.

Philadelphia Inquirer
1878 Obituary. Bloomfield H. Moore. *Philadelphia Inquirer,* July 6.
1918a Teachers to Give Instruction Free. Show Patriotism in Movement to Prepare Women for Skillful War Service. Dr. Wm. P. Wilson Secures Mayor's Support in Organizing Summer Courses. *Philadelphia Inquirer* May 26.
1918b War School for Girls. Institution at William Penn High Opens Tomorrow. *Philadelphia Inquirer* June 30.

Philadelphia Press
1899a Fortune Was Squandered in Keely Motor. *Philadelphia Press* January 22:1, 14.
1899b Keely Secrets Laid Bare. *Philadelphia Press* January 19:1, 6–8.

Philadelphia Record
1915 Woman May Trace Indians' Ancestors to Stone Age. *Philadelphia Record* July 19.
1916 Pueblo Relics Bared by Woman Excavator. *Philadelphia Record* August 31.

Phillips, P.

1973 Introduction. In *Exploration of Ancient Key Dwellers' Remains on the Gulf Coast of Florida,* by Frank Hamilton Cushing, pp. i–xvii. AMS, New York.

Phillips, P., and J. A. Brown

1978 *Pre-Columbian Shell Engravings from the Craig Mound at Spiro, Oklahoma,* Part I. Peabody Museum of Archaeology and Ethnology, Harvard University, Cambridge.

Pinsky, V.

1992 Archaeology, Politics, and Boundary-Formation: The Boas Censure (1919) and the Development of American Archaeology During the Inter-war Years. In *Rediscovering Our Past: Essays on the History of American Archaeology,* edited by J. E. Reyman, pp. 161–89. Aldershot, Avebury.

Powell, J. W.

1883 The Trenton Gravels. *Science* 1:525–26.

1878 The Pueblo Indians. *Rocky Mountain Presbyterian* 7(11):1.

1892 Indian Linguistic Families of America North of Mexico. *Seventh Annual Report of the Bureau of Ethnology, 1884–85,* pp. 1–142.

1890 Problems of American Archaeology. *The Forum* 8:638–52.

1891 The Humanities. *The Forum* 10:410–22.

1892 Remarks on the Classification and Nomenclature of Anthropology. *American Anthropologist,* o.s. 5:266–71.

1900 Anthropology. *American Universal Cyclopedia* 1, pp. 15–17.

Preucel, R. W., and M. S. Chesson

1994 Blue Corn Girls: A Herstory of Three Early Women Archaeologists at Tecolote, New Mexico. In *Women in Archaeology,* edited by C. Claassen, pp. 67–84. University of Pennsylvania Press, Philadelphia.

Public Ledger (Philadelphia)

1895 Mrs. Clara J. Moore Wants Her Son Removed as Co-Executor. *Public Ledger* December 2:16.

1897 Mrs. Bloomfield H. Moore's Answer. *Public Ledger* January 20:13.

1899 Noted Woman Dead. *Public Ledger* January 6:1.

1929 Aide Charges "Secret Sale" of Relics Here. *Public Ledger* May 8:1, 10.

Purdy, B. A.

1996 *Indian Art of Ancient Florida.* University Press of Florida, Gainesville.

Putnam, F. W.
1886a Report of the Curator. *Peabody Museum Annual Report* 18:401–18.
1886b Report of the Curator. *Peabody Museum Annual Report* 19:477–501.
1888 Untitled Remarks, at the December 21, 1887, Symposium "Paleolithic Man in Eastern and Central North America." Proceedings Boston Society of Natural History 23:421–24, 447–49.
1889 Report of the Annual Meeting. *Peabody Museum Annual Report* 23:73–81.

Putnam, F. W et al.
1888 Symposium on "Paleolithic Man in Eastern and Central North America." *Proceedings Boston Society of Natural History* 23:419–49.
1889 Symposium on "Paleolithic Man in Eastern and Central North America. Part III." *Proceedings Boston Society of Natural History* 24:141–65.

Reingold, N.
1991 *Science, American Style.* Rutgers University Press, New Brunswick.

Renfrew, C.
1980 The Great Tradition Versus the Great Divide: Archaeology as Anthropology? *American Journal of Archaeology* 84:287–98.

Reyman, J. E.
1999 Women in Southwestern Archaeology 1895–1945. In *Assembling the Past. Studies in the Professionalization of Archaeology,* edited by A. B. Kehoe and M. B. Emmerichs, pp. 213–28. University of New Mexico Press, Albuquerque.

Riley, T. J.
1987 Ridged-Field Agriculture and the Mississippian Economic Pattern. In Emergent Horticultural Economics of the Eastern Woodlands, edited by W. F. Keegan, pp. 21–27. Occasional Paper No. 7, Center for Archaeological Investigations, Southern Illinois University at Carbondale.

Roberts, D. G.
1999a Memorial: John L. Cotter, 1911–1999. *Historical Archaeology* 33(4):6–18.
1999b A Conversation With John L. Cotter. *Historical Archaeology* 33(2):1–50.

Rossiter, M. W.
1982 *Women Scientists in America: Struggles and Strategies to 1940.* Johns Hopkins Press, Baltimore.

Rouse, I.
1950 *A Survey of Indian River Archeology, Florida.* Yale University Publications in Anthropology 44. New Haven.

Rowe, J. H.
1954 Max Uhle, 1856–1944. A Memoir of the Father of Peruvian Archaeology. *University of California Publications in American Archaeology and Ethnology* 46(1):1–134. Berkeley.

Royal, R.
1992 *1492 and All That: Political Manipulations of History.* Ethics and Public Policy Center, Washington, D.C.

Ryan, J., and C. Sackrey
1984 *Strangers in Paradise.* South End Press, Boston.

Salisbury, R.
1893 Surface Geology: Report of Progress. *Annual Report of the State Geologist for the Year 1892,* 33–166.

Sargent, W.
1799 A Drawing [and accompanying letter] of Some Untensils, or Ornaments, Taken from an Old Indian Grave, at Cincinnati, County of Hamilton, and Territory of the United States, North-West of the River Ohio. *American Philosophical Society Transactions* 4:179–80.

Scarborough, V. L., and D. R. Wilcox (editors)
1991 *The Mesoamerican Ballgame.* University of Arizona Press, Tucson.

Schuyler, R. L.
1999 John L. Cotter Award in Historical Archaeology. *Historical Archaeology* 33(2):1–5.

Scharf, J. T.
1888 *History of Delaware 1609–1888,* vol. 2. L. H. Everts, Philadelphia.

Sears, W. H.
1977 Seaborne Contacts Between Early Cultures in Lower Southeastern United States and Middle Through South America. In *The Sea in the Pre-Columbian World,* edited by E. P. Benson, pp. 1–15. Dumbarton Oaks Research Library and Collections, Harvard University, Washington, D.C.

1982 *Fort Center: An Archaeological Site in the Lake Okeechobee Basin.* University of Florida Presses, Gainesville.

Seed, P.
1995 *Ceremonies of Possession in Europe's Conquest of the New World, 1492–1640.* Cambridge University Press, Cambridge.

Shapin, S.
1996 *The Scientific Revolution.* University of Chicago Press, Chicago.

Sheets-Pyenson, S.
1988 *Cathedrals of Science: The Development of Colonial Natural History Museums during the Late Nineteenth Century.* McGill-Queens University Press, Kingston.

Shipman, P.
1994 *The Evolution of Racism. Human Differences and the Use and Abuse of Science.* Simon and Schuster, New York.

Shippen, P. [S. Y. Stevenson]
1915 Mrs. William P. Wilson, on Archaeological Expedition, to Excavate Indian Ruins. *Philadelphia Record,* July 15.

Short, C. W., and M. D. Plate
1818 Description of an Indian Fort in the Neighborhood of Lexington, Kentucky. *American Philosophical Society Transactions,* n.s. 1:310–12.

Simmons, J. W.
n.d. Manuscript reports. Archives A-43, A-402, and AA-26. Manuscript on file, Arizona State Museum, University of Arizona, Tucson.

Silverberg, R.
1968 *Moundbuilders of Ancient America: The Archaeology of a Myth.* New York Graphic Society, Greenwich.

Snead, J. E.
1999 Science, Commerce, and Control: Patronage and the Development of Anthropological Archaeology in the Americas. *American Anthropologist* 101(2): 256–71.
2001 *Ruins and Rivals, The Making of Southwest Archaeology.* University of Arizona Press, Tucson.

Southall, J.
1878 *The Epoch of the Mammoth.* J. B. Lippincott, Philadelphia.

Squier, E. G.
1849 Aboriginal Monuments of the Mississippi Valley. *Transactions of the American Ethnological Society* 2:131–207.

Squier, E. G., and E. W. Davis
1848 *Ancient Monuments of the Mississippi Valley.* Smithsonian Contributions to Knowledge 1. Washington, D.C.

Spencer-Wood, S. M.
1999 Gendering Power. In *Manifesting Power: Gender and the Interpretation of Power in Archaeology,* edited by T. L. Sweely, pp. 175–83. Routledge, London.

Stanton, W.
1960 *The Leopard's Spots. Scientific Attitudes Toward Race in America, 1815–59.* University of Chicago Press, Chicago.

Starr, F.
1892 Anthropological Work in America. *Popular Science Monthly* 41:289–307.

Stegner, W.
1954 *Beyond the Hundredth Meridian: John Wesley Powell and the Second Opening of the West.* Houghton Mifflin, Boston.

Stevenson, S. Y.
1892a On Certain Symbols Used in the Decoration of Some Potsherds from Daphnae and Naukratis, Now in the Museum of the University of Pennsylvania. *Proceedings of the Numismatic and Antiquarian Society of Philadelphia, 1890–91,* pp. 1–51.
1892b Mr. Petrie's Discoveries at Tel el-Amarna. *Science* 19:480–82, 510.
1893 An Ancient Egyptian Rite Illustrating a Phase of Primitive Thought. *Memoirs of the International Congress of Anthropology,* pp. 298–311.
1894 The Feather and the Wing in Early Mythology. *Oriental Studies. Oriental Club of Philadelphia,* pp. 1–39.
1896 On the Remains of the Foreignors [*sic*] Discovered in Egypt by Mr. W. M. Flinders Petrie, 1895, Now in the Museum of the University of Pennsylvania. *Proceedings of the American Philosophical Society* 35:57–64.
1899 *Maximilian in Mexico: A Woman's Reminiscences of the French Intervention 1862–1867.* New York: Century.

1901 Egypt, Babylonia and Greece. *Report of the Committee on Awards of the World's Columbian Commission. Special Reports upon Special Subjects or Groups* 1:335–46. Washington, D.C.

1907–21 Peggy Shippen's Diary. Weekly column in the *Philadelphia Public Ledger.*

1909 The Training of Curators. *Proceedings of the American Association of Museums* 3:115–19.

Stevenson, S. Y., M. Jastrow Jr., and F. Justi

1905 *Egypt and Western Asia in Antiquity.* Lea Brothers, Philadelphia.

Stewart, K. M.

1967 Excavations at Mesa Grande, A Classic Period Hohokam Site in Arizona. *Masterkey* 41(1):14–25.

Stocking, G. W., Jr.

1987 *Victorian Anthropology.* Free Press, New York.

Stocking, G. W., Jr. (editor)

1996 *Volksgeist as Method and Ethic. Essays on Boasian Ethnography and the German Anthropological Tradition.* University of Wisconsin Press, Madison.

Stoltman, J. B.

1973 The Southeastern United States. In *The Development of North American Archaeology: Essays in the History of Regional Traditions,* edited by J. E. Fitting, pp. 114–50. Pennsylvania State University Press, University Park.

Strouse, J.

1999 *Morgan, American Financier.* Random House, New York.

Sturtevant, W. C.

1969 Does Anthropology Need Museums? *Proceedings of the Biological Society of Washington* 82:619–50.

Swift, F. R.

1903 *Florida Fancies.* G. P. Putnam's Sons, New York.

Taylor, W. W.

1948 *A Study of Archaeology.* American Anthropological Association Memoir 69.

Tebeau, C. W.
1955 *The Story of the Chokoloskee Bay Country with the Reminiscences of Pioneer C. S. "Ted" Smallwood.* University of Miami Press, Miami.

Terrell, J. U.
1969 *The Man Who Rediscovered the West.* Weybright and Talley, New York.

Thomas, G. E., M. J. Lewis, and J. A. Cohen
1991 *Frank Furness: The Complete Works.* Princeton Architectural Press, New York.

Thomas, I.
1820 Origins of the American Antiquarian Society. *Archaeologia Americana, Transactions of the American Antiquarian Society* 1:17–59.

Tichy, M. F.
1941 Six Game Pieces from Otowi. *El Palacio* 48(1):1–6.
1947 A Painted Ceremonial Room at Otowi. *El Palacio* 54(3):58–69.

Tompkins, C.
1970 *Merchants and Masterpieces, the Story of the Metropolitan Museum of Art.* E. P. Dutton, New York.

Topinard, P.
1893 L'anthropologie aux États-Unis. *L'Anthropologie* 4:301–51.

Traubel, H.
1961 *With Walt Whitman in Camden,* vol. 1. Rowman and Littlefield, New York. Originally published in 1905. University Museum Archives. University of Pennsylvania Museum of Archaeology and Anthropology Director's Files. Box 1.

Van Ness, C. M.
1985 The Furness-Mitchell Coterie: Its Role in Philadelphia's Intellectual Life at the Turn of the Twentieth Century. Unpublished Ph.D. Dissertation, University of Pennsylvania.
1988 Sara Yorke Stevenson. *Women Anthropologists: A Biographical Dictionary,* edited by U. Gacs, A. Khan, J. McIntyre, and R. Weinberg, pp. 344–49. Greenwood, New York.

Walthall, J. A.
1980 *Prehistoric Indians of the Southeast: Archaeology of Alabama and the Middle South.* University of Alabama Press, Tuscaloosa.

Wallace, A. R.
1887 The Antiquity of Man in North America. *Nineteenth Century* 22:667–79.

Wardle, H. N.
1929 Wreck of the Archaeological Department of the Academy of Natural Sciences of Philadelphia. *Science* 70(1805):119–21.
1956 Clarence Bloomfield Moore (1852–1936). *Bulletin of the Philadelphia Anthropological Society* 9(2):9–11.

Ware, A. L.
1876 *The First Triennial Report of the Secretary of the Class of 1873 of Harvard College.* Boston.
1888 *The Fifth Triennial Report of the Secretary of the Class of 1873 of Harvard College.* Boston.
1898 *The Seventh Report of the Secretary of the Class of 1873 of Harvard College.* Boston.

Warner, S. B.
1968 *The Private City: Philadelphia in Three Periods of Its Growth.* University of Pennsylvania Press, Philadelphia.

Washington Post
1894 The Pueblos at Home, New Exhibits Being Arranged at the National Museum. *Washington Post,* September 4, 1894.

Whitehill, W. M.
1970 *Museum of Fine Arts, Boston. A Centennial History,* 2 vols. Harvard University Press, Cambridge.

Who Was Who in America
1943 *Who Was Who in America.* Volume 1. 1897–1942. A. N. Marquis, Chicago.

Widmer, R. J.
1989 The Relationship of Ceremonial Artifacts from South Florida with the Southeastern Ceremonial Complex. In *The Southeastern Ceremonial Complex: Artifacts and Analysis,* edited by P. Galloway, pp. 166–80. University of Nebraska Press, Lincoln.

Wilcox, D. R.

1987 *Frank Midvale's Investigation of the Site of La Ciudad.* Arizona State University Archaeological Field Studies No. 19. Tempe.

1999 The Chicago "Triumph" of Frank Hamilton Cushing in 1893. Paper prepared for the symposium, "Coming of Age in Chicago: Studies of Anthropology and Archaeology at the 1893 World's Fair," organized by C. M. Hinsley and D. R. Wilcox. Society for American Archaeology, 64th Annual Meeting, Chicago, March 26, 1999.

2002 Curating Field Anthropology: Why Remembering Matters. In *A History of Anthropology at the Field Museum, 1890–2000,* edited by S. E. Nash and G. Feinman. Fieldiana: Anthropology, in press.

Wilcox, D. R., and D. D. Fowler

2002 The Beginnings of Anthropological Archaeology in the North American Southwest: From Thomas Jefferson to the Pecos Conference. *Journal of the Southwest* 44(2):121–234.

Wilcox, D. R., G. Robertson Jr., and J. S. Wood

2001 Organized for War: The Perry Mesa Settlement System and Its Central Arizona Neighbors. In *Deadly Landscapes: Case Studies in Prehistoric Southwestern Warfare,* edited by G. E. Rice and S. A. LeBlanc, pp. 109–40. University of Utah Press, Salt Lake City.

Wilkins, M.

1970 *The Emergence of Multinational Enterprise: American Business Abroad from the Colonial Era to 1914.* Harvard University Press, Cambridge.

Willey, G. R.

1949 *Archeology of the Florida Gulf Coast.* Smithsonian Miscellaneous Collections 113. Smithsonian Institution, Washington, D.C.

1973 Introduction to Samuel Haven, *Archaeology in the United States,* pp. iv–ix. AMS, published for the Peabody Museum, New York.

Willey, G. R., and J. A. Sabloff

1980 *A History of American Archaeology,* 2nd ed. W. H. Freeman, San Francisco.

1993 *A History of American Archaeology,* 3rd ed. W. H. Freeman, New York.

Williams, R. A., Jr.

1990 *The American Indian in Western Legal Thought.* Oxford University Press, New York.

Williams, S.

1991 *Fantastic Archaeology: The Wild Side of North American Prehistory.* University of Pennsylvania Press, Philadelphia.

Williams, S. (editor)

1977 [1968] *The Waring Papers: The Collected Papers of Antonio J. Waring, Jr.* Peabody Museum, Harvard University, Cambridge.

Williams, S., and J. P. Brain

1983 *Excavations at the Lake George Site, Yazoo County, Mississippi, 1958–1960.* Papers of the Peabody Museum of Archaeology and Ethnology, vol. 74. Harvard University, Cambridge.

Wilson, D. L. (editor)

1967 *The Genteel Tradition, Nine Essays by George Santayana.* Harvard University Press, Cambridge.

Wilson, L. L. W.

1916a A Prehistoric Anthropomorphic Figure from the Rio Grande Basin. *American Anthropologist* 18:548–51.

1916b Excavations at Otowi. *El Palacio* 3(2):28–36.

1917a Excavations at Otowi, New Mexico. *Art and Archaeology* 6:259–60.

1917b This Year's Work at Otowi. *El Palacio* 4(4):87.

1918a Hand Sign or Avanyu. A Note on a Pajaritan Biscuit-Ware Motif. *American Anthropologist* 20(3):310–17.

1918b Three Years at Otowi. *El Palacio* 5:290–95.

Wilson, R. G.

1979 *The American Renaissance, 1876–1917.* Brooklyn Museum and Pantheon, New York.

Wilson, T.

1893 Primitive Industry. *Smithsonian Institution Annual Report 1892*, pp. 521–34.

Winegrad, D. P.

1993 *Through Time, Across Continents: A Hundred Years of Archaeology and Anthropology at the University Museum.* University Museum, University of Pennsylvania, Philadelphia.

Winsor, J. (editor)

1888 *Narrative and Critical History of America.* Houghton, Mifflin, New York.

Wissler, C.
1917 The New Archaeology. *American Museum Journal* 17(2):100–101.
1942 The American Indian and the American Philosophical Society. *American Philosophical Society Proceedings* 86(1):189–204.

Wister, F. A.
1922 *Sara Yorke Stevenson.* Civic Club of Philadelphia, Philadelphia.

Wood, J. S., and D. R. Wilcox
2000 Where Did All the Flowers Go? Warfare and Religion in the Evolution of Hohokam Political Organization. Paper prepared for the Symposium on Hohokam Political Organization, organized by D. R. Abbott and D. R. Wilcox, 65th Annual Meeting of the Society for American Archaeology, Philadelphia, April 7, 2000.

Worster, D.
2001 *A River Running West: The Life of John Wesley Powell.* Oxford University Press, New York.

Wright, G. F.
1889 *The Ice Age in North America and Its Bearing upon the Antiquity of Man.* D. Appleton, New York.
1892 *Man and the Glacial Period.* D. Appleton, New York.

Wyman, J.
1869 Report of the Curator. *Second Annual Report of the Trustees of the Peabody Museum of American Archaeology and Ethnology,* pp. 5–23.
1872 Report of the Curator. *Fifth Annual Report of the Trustees of the Peabody Museum of American Archaeology and Ethnology,* pp. 5–30.

Contributors

Lawrence E. Aten is retired from the National Park Service. He is a Research Fellow at the University of Texas, Austin, and currently is researching the life of Clarence B. Moore (in collaboration with J. T. Milanich) and the archaeology of the Sabine Lake area of Texas and Louisiana. His most recent monograph is "A Late Holocene Settlement in the Taylor Bayou Drainage Basin," coauthored with C. N. Bollich (Texas Archaeological Society Special Publication no. 4, 2002).

Steven Conn teaches cultural and intellectual history at Ohio State University. His first book, *Museums and American Intellectual Life, 1876–1926,* was published in 1998. His most recent book, *Staring at the Past: Native Americans and the Problem of History,* is forthcoming from the University of Chicago Press.

Elin C. Danien is a Research Associate in the American Section of the University of Pennsylvania Museum of Archaeology and Anthropology. She originated and for many years organized the Museum's Annual Maya Weekend. Her most recent publication is *A Guide to the Mesoamerican Gallery at the University of Pennsylvania Museum,* following her renovation of that gallery.

Regna Darnell is Professor of Anthropology and Director of the Centre for Research and Teaching of Canadian Languages at the University of Western Ontario. She is a fellow of the Royal Society of Canada. Her most recent book is *Invisible Genealogies: A History of Americanist Anthropology* (University of Nebraska Press, 2001).

Don D. Fowler is the Mamie Kleberg Professor of Historic Preservation and Anthropology at the University of Nevada, Reno. He is a past-president of the Society for American Archaeology. His most recent book is *A Laboratory for Anthropology: Science and Romanticism in the American Southwest, 1846–1930* (University of New Mexico Press, 2000).

Curtis M. Hinsley is Regent's Professor of History in the Department of Applied Indigenous Studies at Northern Arizona University. He is currently editing, with David R. Wilcox, a multivolume documentary history of the Hemenway Southwestern Archaeological Expedition, 1886–1889. The second volume of the history, *The Lost Itinerary of Frank Hamilton Cushing,* is now available (University of Arizona Press, 2002).

Alice Beck Kehoe is Professor of Anthropology Emerita at Marquette University. Her most recent books are *The Land of Prehistory: A Critical History of American Archaeology* (Routledge, 1998) and *Shamans and Religion: an Anthropological Experiment in Critical Thinking* (Waveland, 2000).

Eleanor M. King is Professor of Anthropology at Howard University in Washington, D.C. She has a long-standing interest in the history of anthropology and has published on the early history of the University of Pennsylvania Museum as well as on the careers of some of its Americanist scholars. A certified archivist, she is co-author of *Preserving Field Records: Archival Techniques for Archaeologists and Anthropologists* (University of Pennsylvania Museum, 1985) and has published articles on the importance of archival management specifically for archaeology.

Frances Joan Mathien, an archaeologist with the National Park Service, serves as general editor of the Chaco Project publications. Her recent contribution is an edited volume, *Ceramics, Lithics, and Ornaments of Chaco Canyon: Analyses of Artifacts from the Chaco Project, 1971–1987* (National Park Service Publications in Archaeology 18G, 1997).

David J. Meltzer is the Henderson/Morrison Professor of Prehistory at Southern Methodist University, Dallas. He has published extensively on Pleistocene peoples and environments of North America and the history of the human antiquity controversy in late nineteenth/early twentieth century America. He is also editor of The Smithsonian Institution's 150th anniversary edition of E. Squier and E. Davis's *Ancient Monuments of the Mississippi Valley* (1998).

Jerald T. Milanich is Curator of Archaeology at the Florida Museum of Natural History. He has published extensively on the precolumbian and colonial period Native American societies of the southeastern United States.

Jeremy A. Sabloff is the Williams Director of the University of Pennsylvania Museum of Archaeology and Anthropology. A member of the National Academy of Science, he is the author of many important works on Maya civilization and the history of Americanist archaeology.

Robert L. Schuyler is Associate Professor of Anthropology and Associate curator in charge of Historical Archaeology at the University of Pennsylvania and its museum. He is a past president of the Society for Historical Archaeology and past executive officer of the Council for Northeast Historical Archaeology. He has conducted archaeological research in New York, New England, and the Greater Southwest. He is currently developing the South Jersey Project, a study of nineteenth and twentieth centuries in the region with a special focus on the town of Vinland.

David R. Wilcox is Senior Curator of Anthropology at the Museum of Northern Arizona, Flagstaff. He is co-editor with Curtis M. Hinsley of a proposed seven-volume work on the Hemenway Southwestern Archaeological Expedition. The second volume of the history, *The Lost Itinerary of Frank Hamilton Cushing,* is now available (University of Arizona Press, 2002).

Index